THE ANNALS

of The American Academy *of* Political
and Social Science

RICHARD D. LAMBERT, *Editor*
ALAN W. HESTON, *Associate Editor*

THE PRIVATE SECURITY INDUSTRY: ISSUES AND TRENDS

Special Editor of this Volume

IRA A. LIPMAN

Chairman and President
Guardsmark, Inc.

SAGE PUBLICATIONS *NEWBURY PARK BEVERLY HILLS LONDON NEW DELHI*

THE ANNALS

© 1988 *by* The American Academy *of* Political *and* Social Science

ERICA GINSBURG, *Assistant Editor*

All rights reserved. No part of this volume may be reproduced or utilized in any form or by any means, electronic or mechanical, including photocopying, recording or by any information storage and retrieval system, without permission in writing from the publisher.

Editorial Office: 3937 Chestnut Street, Philadelphia, Pennsylvania 19104.

For information about membership* (individuals only) and subscriptions (institutions), address:

SAGE PUBLICATIONS, INC.
2111 West Hillcrest Drive   275 South Beverly Drive
Newbury Park, CA 91320   Beverly Hills, CA 90212

From India and South Asia,   *From the UK, Europe, the Middle*
write to: *East and Africa, write to:*
SAGE PUBLICATIONS INDIA Pvt. Ltd.   SAGE PUBLICATIONS LTD
P.O. Box 4215   28 Banner Street
New Delhi 110 048   London EC1Y 8QE
INDIA   ENGLAND

SAGE Production Editors: JANET BROWN and ASTRID VIRDING

*Please note that members of The Academy receive THE ANNALS with their membership.

Library of Congress Catalog Card Number 87-042561
International Standard Serial Number ISSN 0002-7162
International Standard Book Number ISBN 0-8039-3103-4 (Vol. 498, 1988 paper)
International Standard Book Number ISBN 0-8039-3102-6 (Vol. 498, 1988 cloth)
Manufactured in the United States of America. First printing, July 1988.

The articles appearing in THE ANNALS are indexed in *Book Review Index; Public Affairs Information Service Bulletin; Social Sciences Index; Monthly Periodical Index; Current Contents; Behavioral, Social Management Sciences;* and *Combined Retrospective Index Sets.* They are also abstracted and indexed in *ABC Pol Sci, Historical Abstracts, Human Resources Abstracts, Social Sciences Citation Index, United States Political Science Documents, Social Work Research & Abstracts, Peace Research Reviews, Sage Urban Studies Abstracts, International Political Science Abstracts, America: History and Life,* and/or *Family Resources Database.*

Information about membership rates, institutional subscriptions, and back issue prices may be found on the facing page.

Advertising. Current rates and specifications may be obtained by writing to THE ANNALS Advertising and Promotion Manager at the Newbury Park office (address above).

Claims. Claims for undelivered copies must be made no later than three months following month of publication. The publisher will supply missing copies when losses have been sustained in transit and when the reserve stock will permit.

Change of Address. Six weeks' advance notice must be given when notifying of change of address to insure proper identification. Please specify name of journal. Send change of address to: THE ANNALS, c/o Sage Publications, Inc., 2111 West Hillcrest Drive, Newbury Park, CA 91320.

The American Academy of Political and Social Science

3937 Chestnut Street Philadelphia, Pennsylvania 19104

Board of Directors

ELMER B. STAATS	RANDALL M. WHALEY
MARVIN E. WOLFGANG	HENRY W. SAWYER, III
LEE BENSON	WILLIAM T. COLEMAN, Jr.
RICHARD D. LAMBERT	ANTHONY J. SCIRICA
THOMAS L. HUGHES	FREDERICK HELDRING
LLOYD N. CUTLER	

Officers

President
MARVIN E. WOLFGANG

Vice-Presidents
RICHARD D. LAMBERT, *First Vice-President*
STEPHEN B. SWEENEY, *First Vice-President Emeritus*

Secretary	*Treasurer*	*Counsel*
RANDALL M. WHALEY	ELMER B. STAATS	HENRY W. SAWYER, III

Editors, THE ANNALS

RICHARD D. LAMBERT, *Editor* ALAN W. HESTON, *Associate Editor*

THORSTEN SELLIN, *Editor Emeritus*

Assistant to the President
MARY E. HARRIS

Origin and Purpose. The Academy was organized December 14, 1889, to promote the progress of political and social science, especially through publications and meetings. The Academy does not take sides in controverted questions, but seeks to gather and present reliable information to assist the public in forming an intelligent and accurate judgment.

Meetings. The Academy holds an annual meeting in the spring extending over two days.

Publications. THE ANNALS is the bimonthly publication of The Academy. Each issue contains articles on some prominent social or political problem, written at the invitation of the editors. Also, monographs are published from time to time, numbers of which are distributed to pertinent professional organizations. These volumes constitute important reference works on the topics with which they deal, and they are extensively cited by authorities throughout the United States and abroad. The papers presented at the meetings of The Academy are included in THE ANNALS.

Membership. Each member of The Academy receives THE ANNALS and may attend the meetings of The Academy. Membership is open only to individuals. Annual dues: $28.00 for the regular paperbound edition (clothbound, $42.00). Add $9.00 per year for membership outside the U.S.A. Members may also purchase single issues of THE ANNALS for $6.95 each (clothbound, $10.00).

Subscriptions. THE ANNALS (ISSN 0002-7162) is published six times annually—in January, March, May, July, September, and November. Institutions may subscribe to THE ANNALS at the annual rate: $60.00 (clothbound, $78.00). Add $9.00 per year for subscriptions outside the U.S.A. Institutional rates for single issues: $10.00 each (clothbound, $15.00).

Second class postage paid at Philadelphia, Pennsylvania, and at additional mailing offices.

Single issues of THE ANNALS may be obtained by individuals who are not members of The Academy for $7.95 each (clothbound, $15.00). Single issues of THE ANNALS have proven to be excellent supplementary texts for classroom use. Direct inquiries regarding adoptions to THE ANNALS c/o Sage Publications (address below).

All correspondence concerning membership in The Academy, dues renewals, inquiries about membership status, and/or purchase of single issues of THE ANNALS should be sent to THE ANNALS c/o Sage Publications, Inc., 2111 West Hillcrest Drive, Newbury Park, CA 91320. *Please note that orders under $25 must be prepaid.* Sage affiliates in London and India will assist institutional subscribers abroad with regard to orders, claims, and inquiries for both subscriptions and single issues.

THE ANNALS

of The American Academy *of* Political
and Social Science

RICHARD D. LAMBERT, *Editor*
ALAN W. HESTON, *Associate Editor*

──────── FORTHCOMING ────────

CONGRESS AND THE PRESIDENCY:
INVITATION TO STRUGGLE
Special Editor: Roger H. Davidson

Volume 499 September 1988

WHITHER THE AMERICAN EMPIRE:
EXPANSION OR CONTRACTION?
Special Editor: Marvin E. Wolfgang

Volume 500 November 1988

THE GHETTO UNDERCLASS:
SOCIAL SCIENCE PERSPECTIVES
Special Editor: William Julius Wilson

Volume 501 January 1989

See page 3 for information on Academy membership and
purchase of single volumes of **The Annals.**

CONTENTS

PREFACE	*Ira A. Lipman*	9
PRIVATE SECURITY: A RETROSPECTIVE	*Milton Lipson*	11
THE DEVELOPMENT OF THE U.S. SECURITY INDUSTRY	*Robert D. McCrie*	23
PERSONNEL SELECTION IN PRIVATE INDUSTRY: THE ROLE OF SECURITY	*Robert W. Overman*	34
DRUG TESTING IN THE WORKPLACE	*Peter B. Bensinger*	43
EMPLOYEE THEFT: A $40 BILLION INDUSTRY	*Mark Lipman and W. R. McGraw*	51
CIVIL AVIATION: TARGET FOR TERRORISM	*William A. Crenshaw*	60
TECHNOLOGICAL SECURITY	*Felix Pomeranz*	70
PERSONNEL SELECTION IN THE PRIVATE SECURITY INDUSTRY: MORE THAN A RÉSUMÉ	*Ira A. Lipman*	83
THE LEGAL LIABILITY OF A PRIVATE SECURITY GUARD COMPANY FOR THE CRIMINAL ACTS OF THIRD PARTIES: AN OVERVIEW	*Jonathan D. Schiller and Gary K. Harris*	91
THE TIME HAS COME TO ACKNOWLEDGE SECURITY AS A PROFESSION	*Ernest J. Criscuoli, Jr.*	98
CAN POLICE SERVICES BE PRIVATIZED?	*Philip E. Fixler, Jr., and Robert W. Poole, Jr.*	108
BOOK DEPARTMENT		119
INDEX		178

BOOK DEPARTMENT CONTENTS

INTERNATIONAL RELATIONS AND POLITICS

COLLINGRIDGE, DAVID and COLIN REEVE. *Science Speaks to Power: The Role of Experts in Policymaking;* HISKES, ANNE L. and RICHARD P. HISKES. *Science, Technology, and Policy Decisions.* Kenneth P. Ruscio... 119

DORRIEN, GARY J. *The Democratic Socialist Vision.* Willard D. Keim................. 120

FRIEDMANN, JOHN. *Planning in the Public Domain: From Knowledge to Action.* Richard A. Wright.. 121

HUTCHISON, WILLIAM R. *Errand to the World: American Protestant Thought and Foreign Missions.* Dewey D. Wallace, Jr.. 122

LAIRD, ROBBIN F. *The Soviet Union, the West, and the Nuclear Arms Race;* LEBOW, RICHARD NED. *Nuclear Crisis Management: A Dangerous Illusion.* Eric Waldman.. 123

LUTTWAK, EDWARD N. *Strategy: The Logic of War and Peace;* LEVITE, ARIEL. *Intelligence and Strategic Surprises.* William Weida 125

SCHALL, JAMES V. *Reason, Revelation, and the Foundations of Political Philosophy.* Francis M. Wilhoit... 126

AFRICA, ASIA, AND LATIN AMERICA

AMATE, C.O.C. *Inside the OAU: Pan-Africanism in Practice.* Edmond J. Keller 127

BARNES, SANDRA T. *Patrons and Power: Creating a Political Community in Metropolitan Lagos.* Naomi Chazan... 128

BURKI, SHAHID JAVED. *Pakistan: A Nation in the Making.* Theodore P. Wright, Jr. 130

DAVIS, LEONARD. *The Philippines: People, Poverty & Politics.* Gary Hawes 131

HAVENS, THOMAS R. H. *Fire across the Sea: The Vietnam War and Japan 1965-1975.* Edmund S. Wehrle.. 131

KARLSSON, SVANTE. *Oil and the World Order: American Foreign Oil Policy;* KUPCHAN, CHARLES A. *The Persian Gulf and the West: The Dilemmas of Security.* Richard J. Willey 132

MALIK, HAFEZ, ed. *Soviet-American Relations with Pakistan, Iran and Afghanistan;* BENNIGSEN, ALEXANDRE and S. ENDERS WIMBUSH. *Muslims of the Soviet Empire.* Michael Lenker 133

PASTOR, ROBERT A. *Condemned to Repetition: The United States and Nicaragua.* William J. Williams.. 135

RUDOLPH, LLOYD I. and SUSANNE HOEBER RUDOLPH. *In Pursuit of Lakshmi: The Political Economy of the Indian State.* Craig Baxter 136

EUROPE

BRETTELL, CAROLINE B. *Men Who Migrate, Women Who Wait: Population and History in a Portuguese Parish.* Mario D. Zamora..................................... 136

GORDON, LINCOLN with J. F. BROWN, PIERRE HASSNER, JOSEF JOFFE, and EDWINA MORETON. *Eroding Empire: Western Relations with Eastern Europe.* Philip B. Taylor, Jr.. 137

KOONZ, CLAUDIA. *Mothers in the Fatherland: Women, the Family, and Nazi Politics.* Jean Bethke Elshtain 139

ROSE, RICHARD. *Ministers and Ministries: A Functional Analysis.* Henry R. Winkler 140

SYMONDS, RICHARD. *Oxford and Empire: The Last Lost Cause?* Wesley K. Wark 141

UNITED STATES

CAIN, BRUCE, JOHN FEREJOHN, and MORRIS FIORINA. *The Personal Vote: Constituency Service and Electoral Independence.* Darrell M. West 142

COSGROVE, RICHARD A. *Our Lady the Common Law: An Anglo-American Legal Community, 1870-1930.* Karl H. Van D'Elden 143

FERMAN, BARBARA. *Governing the Ungovernable City: Political Skill, Leadership, and the Modern Mayor.* Bruce Edw. Caswell............................. 144

HIGGS, ROBERT A. *Crisis and Leviathan: Critical Episodes in the Growth of American Government.* Rush Welter 145

KUGLER, ISRAEL. *From Ladies to Women: The Organized Struggle for Woman's Rights in the Reconstruction Era.* Robert L. Daniel 146

LEWIS, JOHN S. and RUTH A. LEWIS. *Space Resources: Breaking the Bonds of Earth.* Zachary A. Smith 146

LOWENSTEIN, SHARON R. *Token Refuge: The Story of the Jewish Refugee Shelter at Oswego, 1944-1946.* Hans Segal.. 147

PETERSON, PAUL, BARRY RABE, and KENNETH WONG. *When Federalism Works.* James W. Fossett 148

RIKER, WILLIAM H. *The Development of American Federalism.* Joseph F. Zimmerman 150

ROBINSON, DONALD. *"To The Best of My Ability": The Presidency and the Constitution.* Martin L. Fausold 150

SIMONTON, DEAN KEITH. *Why Presidents Succeed: A Political Psychology of Leadership.* Stephen W. White 151

THERNSTROM, ABIGAIL M. *Whose Votes Count? Affirmative Action and Minority Voting Rights.* Joseph F. Zimmerman 152

SOCIOLOGY

CLAVEL, PIERRE. *The Progressive City: Planning and Participation, 1969-1984.* Ira Harkavy 153

ESTRICH, SUSAN. *Real Rape.* Leslie Lebowitz... 154

LANDRY, BART. *The New Black Middle Class.* David J. Garrow 156

TRAUTMANN, THOMAS R. *Lewis Henry Morgan and the Invention of Kinship.* Karl A. Peter 156

WHITTAKER, ELVI. *The Mainland Haole: The White Experience in Hawaii.* Jeffrey L. Crane 157

ECONOMICS

AMY, DOUGLAS J. *The Politics of Environmental Mediation;* BOSSO, CHRISTOPHER J. *Pesticides and Politics: The Life Cycle of a Public Issue.* Duane Windsor 159

ARNDT, A. W. *Economic Development: The History of an Idea.*
Winston E. Langley 160

GALENSON, DAVID W. *Traders, Planters, and Slaves: Market Behavior in Early English America.* Jan Hogendorn 161

HIBBS, DOUGLAS A. *The American Political Economy: Macroeconomic and Electoral Politics in the United States.* Kenneth K. Wong 162

JENKINS, RHYS. *Transnational Corporations and the Latin American Automobile Industry.* Mira Wilkins 163

MARSHALL, RAY. *Unheard Voices: Labor and Economic Policy in a Competitive World;* FLANAGAN, ROBERT J. *Labor Relations and the Litigation Explosion.*
Duncan Colin Campbell 164

PAUL, ELLEN FRANKEL. *Property Rights and Eminent Domain.* Wallace F. Smith 166

RIDDELL, ROGER C. *Foreign Aid Reconsidered.* Wilfred Malenbaum 167

PREFACE

This issue of *The Annals* brings together articles pertaining to both the history and the future of the private security industry in the United States. The private security industry has long suffered under the misconception that it provides the doddering retired military man with the job of guarding a warehouse—while he takes advantage of the opportunity to catch up on some much needed sleep. This profile of the security guard, or night watchman, as he was called, while once somewhat accurate, has undergone changes that affect the present and the future of the industry. In progressive companies, today's security officer may be male or female; he or she may be 25 years old, or perhaps 50; he or she is probably a high school graduate and is increasingly more likely to have some college education, or, in our company's case, nearly 25 percent of the security officers have graduated from college or have attended college for at least one year. Today's security officer is not merely looking for a temporary night job that will furnish the chance to watch television and rest up for the more important day job; today's security officer is choosing security as a career.

The responsibilities of the security officer have changed, and they continue to change, given the dynamic nature of the industry. With police budgets steadily shrinking and the crime rate on the increase, society is feeling the effects of understaffed law enforcement agencies. Many traditional responsibilities of law enforcement agencies are being privatized—including public building security, residential neighborhood patrol, traffic control, parking enforcement, crowd control, and court security—and communities are discovering that their police departments, for instance, can work in concert with the private security industry to provide comprehensive security in a more cost-effective manner than they otherwise could have done.

Public law enforcement agencies are not alone, however, in facing the challenge of providing sufficient security despite dwindling resources. Corporations, too, are anticipating lean times, based upon economists' predictions for an upcoming recession. Some have already begun implementing across-the-board cutbacks in areas such as security and, unfortunately, have subsequently realized that security is perhaps the one area wherein reduction in personnel is a mistake with far-reaching implications. Corporations' employees and former employees have reacted bitterly to their termination; as a result, the businesses have experienced increased loss, particularly of property and information. Again, private security is the efficient solution: corporations can prevent loss without the expense of their own full-scale proprietary security department. Their management can join with an outside private security company for a comprehensive loss prevention program.

Clearly, the security industry will continue to play a more active and important role than ever in the coming years. The public's acceptance of private security officers for what they have become or are becoming—responsible agents in overall

NOTE: The fine work of those who assisted with this issue of *The Annals* was greatly enhanced by the efforts of Susanne Loftis, my associate at Guardsmark, Inc. I am grateful to all those who made this project possible.

public and private loss prevention—rests, however, with the industry itself. Those private security companies that are proven responsible providers of their service must remain examples for the rest of the companies in the industry, some of which are only beginning to respond, for instance, by imposing higher standards for their officers, enforcing those standards through more thorough preemployment screening, and disarming their officers.

In his article in this *Annals* issue, Mr. Lipson provides a retrospective of the security industry from its earliest origins, and Mr. McCrie elaborates on Mr. Lipson's article with an overview of the industry, specifically its three component parts: alarm, armored car, and security guard services.

Security considerations must extend beyond the hardware of the alarm and the armored car, though, to the personnel hired for nonsecurity positions, as Dr. Overman reports. Although preemployment investigation reports on prospective employees can meet with legal roadblocks, it is the responsibility of the human resources manager to conduct or oversee as thorough an investigation as possible, if the corporation expects to have a quality work force. Many companies are now including a drug test among their battery of preemployment tests. In his article, Mr. Bensinger examines the legality of drug tests as well as the types of tests available and their value. Mr. Mark Lipman and Mr. McGraw examine a possible consequence of haphazard employee selection: theft by employees. Theft of technology, specifically through the use of computers, is the subject of Mr. Pomeranz's article. Dr. Crenshaw reports that another result of the involvement of private security is the protection of the domestic air transportation system. Employee theft, computer fraud, theft of information, and attacks on airports and aircraft represent but several of the many areas that private security can positively influence.

For a private security company to provide excellence in service to its clients, it must be certain that it employs only the most honest, trustworthy, law-abiding, and psychologically stable people. Our article examines the ways to guarantee that only these people are hired and what would happen to the security industry if access to these tests and investigations in the personnel selection process were denied. Mr. Schiller and Mr. Harris describe the extent of security companies' liability for the criminal actions of third parties.

Mr. Criscuoli explains that, contrary to the perception of a segment of the public, the field of security does require specialized knowledge and skills and, therefore, should be considered a profession. Dr. Fixler and Mr. Poole report that, although some people still consider police services a public good that only government should provide, several communities have successfully privatized these services.

Private security has made tremendous advances over the last few decades, and this progress is becoming recognized by people outside the industry, as is evidenced by the broad range of disciplines represented in this volume. Doubtless, the industry's expansion into areas once considered the exclusive domain of public law enforcement agencies will evoke even more public confidence in the ability of private security to succeed in nontraditional settings. As we move toward the 1990s, and the twenty-first century, private security, by assuming traditional law enforcement responsibilities, will, more than ever, be an integral part of society.

IRA A. LIPMAN

Private Security: A Retrospective

By MILTON LIPSON

ABSTRACT: The short review of the little-known and little-understood history of private security covers the highlights of the development of police, both public and private, from the beginning—when shepherds were used to guard flocks and warn of raids by other tribes—to the modern period in the United States. Highlighted are the vestigial use of mercenaries by small principalities; the decision by the American Constitutional Convention not to use the French or Continental police system; as well as the manner in which private security, especially as spearheaded by Allan Pinkerton, filled the void in law enforcement in the second half of the nineteenth century.

Milton Lipson has served as investigator on the staff of Congressman F. H. LaGuardia, an agent with the U.S. Secret Service, a practicing attorney, commissioner of investigations for Nassau County, New York, vice-president for corporate security at American Express, adjunct associate professor of security management at New York University, and an assistant district attorney. He is author of On Guard: The Business of Private Security *(1975). He attended Columbia University and earned his LL.B. at St. Lawrence University. He was elected village justice of Sea Cliff, New York, in 1986.*

A familiar figure in our cities, in factories, in stores, at airports, and along docks is the security guard and his guard dog. This aspect of private security is listed in classified telephone directories under "K-9 officers" or "commercial guard dogs" or just "guard dogs." It is just one of the many services available from purveyors of private security to business.

Private security originated in that clouded time when man began to domesticate animals and graze his herds. To safeguard these from both human and animal marauders and to keep them from wandering, one or more of the clan would act as a guardian, a security guard, a shepherd. In time, he was joined by a dog who acted as his valued assistant. The march of the centuries has not materially changed one of the earliest methods of security.

In simple fashion, the foregoing demonstrates the continuing need over the centuries of the services we now call private security. It was among the earliest human occupations, and its evolution to its present place is of interest.

ANCIENT ORIGINS

Nomadic tribes used guards to protect themselves and their flocks from carnivorous beasts and the perils of other raiding tribes. Primitive rules of clan and kinship were born, and retribution became a way of life. Blood feuds, "an eye for an eye," and revenge gave rise to a need for a striking force. Men were armed and formed into military groups that performed both security and attack functions. To this day, our military forces retain the duty of being the ultimate reserve in the function of peacekeeping.

The ancient forces assumed the responsibility of protecting the ruler and his treasures. The names of some of these forces echo still. The Egyptians had their Mamlukes, the Romans their Praetorian Guard, and the sultans were watched over by the Janissaries.

Trade developed between tribes as they grew to nations. Products that were transported by caravan and ship required security. References to this security are found in ancient writings and the Bible. Early codes of law made reference to security officers. The conquests of the regions of Asia Minor by the Greeks under Alexander the Great, the Babylonians, and later the Romans saw the function of public security as a duty of armed forces. In addition to keeping the peace at home, these forces were involved with the subjugation and control of the conquered.

The fall of Rome to the barbarians of the north saw its elaborate systems of law, the military, and commerce replaced by chaos. In the east, the governments of Constantinople continued until replaced by the rulers of Islam. Under the Byzantine emperors, and later under the sultans, the rule of law and a more ordered existence continued. Here private security forces in the form of guards, who were sometimes eunuchs, came into being to protect the wealthy, their wives, and their property.

In Europe, the Dark Ages saw a very gradual return to systems of law and order. In its growth, two separate systems developed. On the European continent, law was based on a civil case environment. Codified by Rome and Constantinople and by the French under Napoleon, these rules form the basis of present-day law. In England, a different system, common law, evolved. These two systems gave rise to variances in the emergence of both public and private police.

THE MIDDLE AGES TO THE VICTORIAN ERA

Despite the emergence of Christianity, the fall of Roman rule saw a resurgence of the primitive rules of kinship, blood feuds, and revenge. Clans entrusted their security to men who were at the same time soldiers, police, and private guards. This continued when the clans banded together to form nations with kings, dukes, lords, and clergy who set their rules of conduct and devised ways to enforce them.

England, prior to the Norman Conquest of 1066, had evolved a system of Tens with each member bound by a frank pledge for the good behavior of the others, answerable in blood or other damages. The Tens similarly were bound together in groups of ten, called Hundreds. There emerged a requirement to keep the king's peace. The old territories of the clans called shires had officers of the king appointed to oversee this peace. Called reeves, each became the shire-reeve, later shortened to "sheriff." This system was kept by the new masters under William of Normandy, who added another officer, called a constable, to patrol the Hundred.

The England that followed saw the rise of great barons and lords, each with his own retainers, often armed, who performed the security function both as public as well as private police. A function that came into its own in hamlets built of wood was that of the fire watch or the night watch. Although fire was needed for heat, light, and cooking, an errant ember could endanger an entire community. This duty soon included that of security watchmen for the castle and the village as well.

By the end of the thirteenth century, by royal decree, the number of constables per Hundred was increased to two. Each citizen was required to join in to catch those who broke the law; failure to apprehend meant punishment. In the cities, standing watch was a duty, along with the others, that growing guilds used in order to regulate the master and apprentice system.

Not long afterward, the French king, Philip I, started paying royal officers assigned to keeping law and order and gathering intelligence. This French system grew into the military police with nationwide power and jurisdiction. By the middle of the eighteenth century, they were a national police force called Gendarmerie, organized down to agents on almost every block. The word "dossier" took on a special meaning, with a dossier maintained on almost every subject. Prostitutes and brothel keepers kept detailed records for the police. Under this setup, there was no room for private security. It was much cheaper to bribe the regular police than to pay retainers of one's own. After the French Revolution, the organization came under the control of Fouché, who remained at its helm until the Battle of Waterloo in 1815. Under him, the organization surpassed in thoroughness the comparable force of the ancien régime.

Smaller countries across Europe followed the example of the French in setting up their own security systems, at whose heart was the use of the military to keep law and order and secure the safety of the ruler. Some developed these systems into exportable surpluses and went into a rent-a-regiment business. The Hessian regiments hired by the English to assist them in their attempt to put down the Americans in the revolution of 1775-81 are but one example. Another that continues to this day is the use by the Vatican of the Swiss Guards. Their use by the Vatican predates the painting of the Sistine Chapel by Michelangelo. It was he who designed the uniforms they still wear while on duty. They represent the oldest continuing guard service.

The period following the Civil War in England, beginning with the restoration of the Stuart kings in 1660 and continuing until the end of the eighteenth century, was marked by a lawlessness unknown in previous history.[1] Detailed in the literature of the times were songs and stories of highwaymen like Dick Turpin. The *Beggars Opera* is one such detailed story of the underworld of the English capital.

Henry Fielding's novel *Tom Jones* has accurate descriptions of that lawless period. Fielding, a particularly good reporter, viewed that scene from the bench as magistrate at Bow Street. He was succeeded in that post by his brother John, and the two left their mark as instigators of reform. Their court officers, known as Bow Street Runners, were probably the most professional of all the law enforcement officers of that era.

The well-organized London gangs were sought out in their known haunts by their victims, who came to buy back their own property. This type of flagrant organized theft reached its zenith under Jonathan Wild, who made his offices in the street of the Old Bailey. When Parliament passed laws aimed at his fencing operations, he turned and advertised that, for an agreed-upon reward and the receipt of a detailed description of the property in question, he would undertake to locate it and restore it to its owners. This ploy also became the target of still another parliamentary enactment. Only then and after three trials was Wild eventually convicted and hanged in 1725. His henchmen had included not only the thieves of London but also many of its constables, security guards, lawyers, servants of the rich, and the like.[2]

By the start of the nineteenth century, it was estimated that 1 out of every 22 people living in England was a professional criminal.[3] Change started not long thereafter. It was due to a combination of many factors, not the least of which was the rapid and extensive growth of laws relating to theft.[4] These new laws helped set the state for the passage of the laws establishing what later became London's Metropolitan Police.

In America, the settlement of newly discovered lands gave rise to different security needs, those of frontiers and Indians. Frontier life promoted the ideas of self-help and mutual aid. Settlers went into their fields with their tools and also their muskets. Their shots and the cannon shot from the nearby stockade warned of danger to all within their sound. The system moved with the frontier on its way west.

The growth of the colonies and the rise of its towns were marked in Massachusetts in 1699 with the creation of a formal night watch. Other areas followed suit. Where slavery flourished, it was economically important that these valuable chattels did not run off. To meet this challenge, plantation owners and

1. John Wade, *A Treatise on Police and Crimes of the Metropolis* (1829; reprint ed., Montclair, NJ: Patterson Smith, 1972), p. 78.

2. W. L. Melville-Lee, *A History of Police in England* (1901; reprint ed., Montclair, NJ: Patterson Smith, 1971), p. 200; Jerome Hall, *Theft, Law and Society*, 2nd ed. (New York: Bobbs-Merrill, 1952), p. 73.

3. *Encyclopedia Britannica*, 14th ed., s.v. "Police."

4. Hall, *Theft, Law and Society*, p. 77.

their overseers organized cooperative security patrols.

In the transition to an independent nation, a decision was made not to follow the French type of national police, with its ramifications. There is no doubt but that these were known to many who exercised influence on the formation of the new government; Benjamin Franklin, John Adams, and Thomas Jefferson, among others, had spent many years at the French court prior to the convening of the Constitutional Convention. Statehood and its quasi-autonomy also militated against the creation of a national police.

Common law that was already established continued after independence. Criminal laws that existed prior to the American Revolution were continued with adjustments made for their new position. American legislatures and lawyers maintained their active interest in the changing criminal laws and rulings of England, particularly the growth of law on theft that came into being at the end of the eighteenth century. Statutes were adopted in most of the new states that followed the English enactments almost without change.[5]

The first quarter of the nineteenth century saw the growth of cities in the New World. Each had its own version of a night watch, and some had a newly organized day force called a ward. These were consolidated and organized into formal police organizations in the period following 1844, when New York legislated day and night police for the city. The next decade witnessed creation of similar departments in such cities as Boston, Chicago, Philadelphia, Baltimore, New Orleans, Saint Louis, Providence, and Newark.

5. Ibid., p. 58.

THE ERA OF PUBLIC POLICE

Societies in the English-speaking countries on both sides of the Atlantic adapted themselves to their new police environments. Helpful in this adaptation was the advent of the Victorian era, with its stricter rules of public morality. There was a greater reliance on the new public police to protect both persons and property, with a diminishing use of private guards and watchmen, especially in Great Britain.

In the United States, expansion westward was the great movement during the nineteenth century. Following the War of 1812 with England, that movement intensified along natural routes, the rivers, and the lakes and was furthered by the building of canals. Along newly opened routes, people and both local and foreign goods moved in increasing volume. Goods needed protection, and watchmen manned docks, barges, and other installations vital to this traffic. Soon thereafter, the developing railroads followed suit with private security of their own.

The growing public police were created by and acted under state and local laws. As now, there was then no national police force in the United States. Nor were there any federal investigative services with nationwide jurisdiction. The Customs Service alone had nationwide responsibility but was confined mainly to the major ports and borders.

Where slavery was legal, the Missouri Compromise of 1820 drew a line across the country limiting its spread. The problem of securing the valuable property represented by a slave increased manyfold by the activities of the Abolitionists who sponsored the escape of slaves and aided and sheltered them on their underground route north to Canada and freedom. To counter this, the large slaveholders pursued their rights in the Congress and the courts and obtained the passage of fugitive slave laws. These were backed up by employment of private investigators in the North to run down and return fugitives. It is conjectured that some investigators so employed later formed the basis of the professional spy network of the Confederate government during the Civil War.

Other responses to the need for private security in that period included the beginnings of express companies engaged in the handling and transportation of valuables. The American Express Company was formally started in 1850 by Henry Wells and Walter Fargo.[6] Their original business was the secure movement of specie and bank documents across New York State from Buffalo to New York City. Two years later, Wells and Fargo expanded into California, where they felt a need existed for similar services in the newly discovered goldfields. Retaining their positions with American Express, they formed a new company bearing their names.

With the opening of the railroads, the express companies began to send valuables in safes accompanied by messengers in their cars. By the end of the 1850s, both Adams Express and American Express were designing and using their own express cars on the railroads. In many respects, the security needs of the railroads and the express companies coincided. Little security assistance was available from the new local police, who lacked jurisdiction away from their own limited areas as well as the necessary financial support for efforts beyond their taxpayers' borders. No federal service existed to which they could turn for help.

This void could not wait for legislation. Pulled into the breach were the newly emerging private security firms. Preeminent among them was the company started by Allan Pinkerton. The history of the company and its founder is illustrative of how private security filled the gap until, much later, federal and state legislation and organizations took over.

Allan Pinkerton, a Scotsman born in 1819, emigrated to Dundee, Kane County, Illinois, some 38 miles from Chicago. He was then 23 years old and had four years' experience as a cooper. A strange coincidence led him into a different venture. While searching in a wooded area for logs suitable for barrel staves, he chanced upon evidence of a plant making counterfeit currency. No paper currency was then issued by the United States government; paper money was issued instead by state banks. Pinkerton contacted both the bank involved and the local sheriff. His efforts resulted in the capture of the counterfeiters and their plant. Pinkerton earned a substantial reward. Within a short time, Pinkerton was made a deputy sheriff of Kane County. Later, he switched to a similar post in Cook County and, in 1849, was appointed to be Chicago's first detective.

The following year, he set himself up as a private investigator. The word "private" on his letterhead in no way limited his great talent for publicity. From the very beginning of his career, he was the subject of numerous newspaper stories. One story said that he had no superior in the country as a detective, and the writer doubted that he had an equal.[7] *The Police Gazette*, first published in 1845, carried regular stories

---

6. Alden Hatch, *American Express* (New York: Doubleday, 1950), p. 15.

7. James D. Horan, *The Pinkertons* (New York: Crown, 1967), p. 24.

about him and his operatives. He later was the author of 18 books about himself and his cases.[8] These books enjoyed large sales and were compared to the memoirs of the French detective François Eugène Vidocq. Vidocq, who started as police informer and then a policeman, rose through the ranks to become second in command of all the French police under Fouché. His memoirs, published in France in 1829, were translated into English and became a best-seller in America.

Pinkerton's agency, thanks to his acumen and the needs of the times, was a success. By 1853, he had a staff of five full-time detectives, one a woman. In 1855, one of his clients, on a retainer of $10,000 a year, was the Illinois Central Railroad.[9] Other railroads that used his services at that time included the Michigan Central; Michigan Southern and Northern Indiana; Chicago and Rock Island; and the Chicago, Burlington, and Quincy. Each of these railroad clients issued free passes over their lines to Pinkerton employees. Another client was the United States Post Office. Eastern railroads, including the Pennsylvania Railroad, joined the group later.

8. Allan Pinkerton's books are the following, all published by G. W. Carleton, New York, between 1870 and 1884: *The Gypsies and the Detectives; A Double Life and the Detectives; Bucholz and the Detectives; Claude Melnotte as a Detective; The Spiritualists and the Detectives; The Mississippi Outlaws and the Detectives; Strikers, Communists, Tramps and Detectives; The Spy of the Rebellion; The Bank-Robbers and the Detectives; The Rail-Road Forger and the Detectives; Criminal Reminiscences and Detective Sketches; The Expressman and the Detectives; The Somnambulist and the Detectives; The Model Town and the Detectives; The Burglar's Fate and the Detectives; The Molly Maguires and the Detectives; Professional Thieves and the Detectives;* and *Thirty Years a Detective.*
9. Horan, *Pinkertons*, p. 31.

The long arm of history and coincidence propelled Pinkerton onto the national scene. George B. McClellan, an engineering graduate of West Point, had, on assignment from the U.S. Army, been its official observer with the British Army in the Crimea. After his return, he resigned his commission and, starting in 1857, became vice-president and chief engineer of the Illinois Central Railroad. In that post, he worked closely with Allan Pinkerton. The railroad's attorney at the state capital in Springfield was Abraham Lincoln, who met with Pinkerton on business on occasion. McClellan later became, by appointment of President Lincoln, the commander in chief of the Union army, and, still later, the 1864 presidential candidate of the Democratic Party defeated by Lincoln.

In February 1861, Lincoln, who had been elected president the previous November, was en route from Springfield to Washington to his inaugural, scheduled for 4 March. His route and schedule provided many stopovers for political consultations. One stopover was to be at Harrisburg, Pennsylvania, from where he was to proceed to Washington by railroad via Baltimore. Baltimore was the southern terminus of the northern railroads and the northern one for railroads of the south. Their terminals, however, were quite a distance apart. Passengers going in either direction had to get from one terminal to the other by carriage or by foot.

Alerted by Pennsylvania Railroad officials, Pinkerton checked out rumors of a plot to assassinate the president-elect as he proceeded from one terminal to the other on his way through Baltimore. Baltimore was, at the time, a city in turmoil, torn between its Northern and Southern sympathizers. Pinkerton's agents confirmed the rumors.

Accompanied by the president of the Pennsylvania Railroad, Pinkerton met with the Lincoln party in Harrisburg and passed on the information. As a result, the president-elect accompanied only by his secretary John Nicolay, Pinkerton, and one of Pinkerton's operatives proceeded to Washington via Baltimore without incident on a train other than the one announced. In its telling, the story became embellished so that it told of Lincoln sneaking past Baltimore's Southern zealots disguised as a woman.

With Lincoln's call for 75,000 volunteers for the Union army, trained officers were in great demand. George McClellan was commissioned a general and given command of the Union Department of Ohio. He appointed his detective friend, Pinkerton, his chief of intelligence.

Pinkerton moved with McClellan to the Army of the Potomac in full charge of all intelligence matters there. He is credited with the establishment of the North's system of spies in the Confederacy, which was created around the nucleus of the old Pinkerton Agency, using its agents, informants, and contacts.

Historical hindsight indicates that McClellan, an admirable organizer and trainer of armies, avoided combat under the exaggerated impression that the size of General Lee's force was far larger than it actually was. This miscalculation was especially costly after the battle of Antietam, where aggressive action by the North could have materially shortened the Civil War. Some of the blame for this misinformation must be put at Pinkerton's door. When Lincoln relieved McClellan of his command in 1863, Pinkerton did not continue with the Union forces but returned to his prewar detective business.

Before taking on the McCellan assignment, Pinkerton's staff had grown to 15 full-time investigators. His return to private work, with the added prestige of his position in the intelligence hierarchy, was marked with still greater success, and, by 1866, he had opened new offices in both New York and Philadelphia. This success was marked by the usual publicity. Before the war, he had solved a case for Adams Express involving a missing package of $40,000. He proved the thief to be the Adams manager in Augusta, Georgia, and recovered the money. The new well-publicized case in 1866 involved a loss to the same company of some $700,000 in cash, bonds, and jewelry taken from a locked safe on a railroad car en route from Rye, New York, to New Haven, Connecticut. Dogged detailed investigation disclosed that the loot must have been thrown from the train while it was in motion. Careful searching along miles of track turned up a bag containing $5000 in coins in a patch of weeds near the right-of-way. This recovery led first to a railroad brakeman and then to the entire gang. A great portion of the loot was also recovered.

Labor strife in the age of expansion

After the Civil War, America entered one of its great periods of industrial expansion. At the same time, large portions of the Midwest and West were being settled. There was a great demand in both areas for manpower. Profitably filling this need, the railroads and steamship lines flourished by bringing emigrants to this country from all parts of Europe. Many were steered into jobs in mining and the emerging heavy industries. Employment conditions and local agitators combined to give rise to labor unrest and eventually to labor union movements of various political and social persuasions. To obtain necessary intelligence to combat these groups, management turned to trusted employees and to the private security companies. Pinkerton's was in the forefront of those so retained.[10]

An underground organization called the Molly Maguires, a violent group, became a major problem for company owners. Pinkerton, in the employ of the owners, placed one of his own agents, James McPartland, in the gang, from which, from 1873 to 1876, he reported their movements. He surfaced to testify against their leaders, whose criminal convictions effectively destroyed that movement. Here again, the resultant publicity increased the demand for private security.

For the next quarter century and more, management and labor engaged in industrial strife of no mean proportion. During periods of strikes, it was not unusual for episodes of brick throwing, window breaking, fighting, shooting, and the use of dynamite to be attributed to labor. Against this, management used an army of mercenaries. These were so-called toughs, enlisted for relatively high pay by private security firms for the duration of a local strike. Again, the Pinkerton organization was in the forefront.

What can be designated only as a battle involved the confrontation of July 1892 at the Homestead Works of the Carnegie Steel Company on the banks of the Monongahela River not far from Pittsburgh, Pennsylvania. On one side were the members of the Amalgamated Association of Iron and Steel Workers, successors of the Son of Vulcan, the skilled workers at the plant. The company had come under the management of Henry Frick, standing in for Andrew Carnegie, who had gone into semiretirement in Scotland. The union represented but 800 of the plant's 3800 employees. A deadlock occurred in negotiation as to pay, Frick demanding that a new contract include a cut in pay from $25 per week to $23 and the union reluctantly saying they would accept $24.

Like many other such installations, the Homestead plant was surrounded by a strong fence topped with barbed wire. Strategically placed along the perimeter were elevated platforms equipped with searchlights. To man these, Frick called upon the Pinkerton organization. It was the seventy-third time that Pinkerton was hired by management in such a situation. In each such case, the local sheriff was persuaded that a posse was needed to preserve law and order.[11] Because the sheriff was empowered to impress private citizens into such a posse, it followed that he could so enlist the employees of Pinkerton without the need to call upon reluctant local residents.

In June 1892, Frick contracted with Pinkerton to supply 300 men at $5 per day to be sent to Pittsburgh. Pinkerton followed up by shipping 250 Winchester rifles and 300 pistols to the Union Supply Company, a Carnegie subsidiary in Pittsburgh. Local attorneys acting for management contacted the sheriff of Allegheny County, who promised to deputize the Pinkerton men after they had actually taken over the custody of the Homestead Works.

On 1 July, when no agreement was reached, the labor union's men occupied the plant and its perimeter. In addition, the union set up barricades at the railroad

10. Burton J. Hendrick, *The Life of Andrew Carnegie*, 2 vols. (New York: Doubleday, Doran, 1932), 1:389.

11. Ibid.

station and across roads leading to the plant. They also set up an elaborate warning system to let them know of any group's approach. The mayor of Homestead was both a plant employee and a striker, so no attempt was made by management to gain his assistance. At 2:00 a.m. on 6 July, 300 armed Pinkerton men embarked on a two-hour journey by river barge to the river entrance to the Homestead plant. This entrance was supposed to be the weak link in the plant's defenses.

The labor union's warning system worked, and the Pinkerton men were met by a reception committee estimated by some to be as many as 10,000. The tug maneuvered the barges alongside the plant and then abandoned them there when both sides opened fire. Eight men were killed that morning, five strikers and three Pinkertons.

Abandoned and outnumbered, the Pinkerton men surrendered and were escorted through a gauntlet to the railroad station. According to Burton J. Hendrick, the authorized biographer of Andrew Carnegie,

> The chief offenders were women, the wives of Hungarians, Slavs and Italians; the cowardly Amazons lining both sides of the advancing procession, beat the unarmed men with clubs, hurled stones and pieces of iron, until the march was changed into a mass of stumbling, falling, half-crazed, bleeding men.

Four days later, without any opposition, the plant was occupied by troops called out by the governor. Two weeks later, Henry Frick was shot and seriously wounded as he sat at his desk in Pittsburgh.

All this occurred during the campaign for president of the United States by incumbent Benjamin Harrison running against the man he had defeated four years before, Grover Cleveland. The strike and what happened thereafter promptly became campaign issues. A full-scale investigation was conducted by the House of Representatives. The union, when it failed to settle the strike, disbanded. Pinkerton announced that supplying "watchmen" in labor disputes was dangerous and that it would no longer engage in that practice. The following year, Congress enacted the so-called Pinkerton Law, which barred Pinkerton's and similar agencies from employment by the U.S. government.[12]

The settlement of the West in the post-Civil War period required great use of private security. This need emerged from such common occurrences as claim jumping, cattle rustling, horse stealing, and coach and train robbery, as well as all the other hazards of newly established areas. Localities were settled long before the arrival of troops or federal or territorial officials. Vigilante groups, formed in the heat of exasperation with being victimized in what ended as summary necktie parties, did little to help in the long run. Sworn law officers existed in some areas but had limited jurisdiction and resources. Few crimes came under federal jurisdiction. For example, train robbery was not legislated to be a federal crime until after World War I.[13] Federal officials were few and virtually powerless.

Valuables were entrusted to such agencies as Adams Express, Wells Fargo, Overland Express, and others. Banks and mercantile establishments sprang

---

12. 27 Stat. 591, 5 U.S.C. 53, enacted 3 March 1893, reads in part as follows: "Hereafter no employee of the Pinkerton Detective Agency, or similar agency, shall be employed in any Government Service, or by any officer of the District of Columbia." Revised in 1966 by Pub. L. 89-554, now cited as 5 U.S.C.A. § 3108.

13. Horan, *Pinkertons*, p. 262.

up, all offering security as part of their service. This security took the form of in-house or contract security guards and detectives.

The Pinkerton contribution to this period is easiest to trace. That company, the first to do over $1 million in business a year—achieved in 1868-69—opened a branch in Denver. In the ensuing period, it broke up the Reno gang of train robbers, chased the James brothers, Jesse and Frank, and were long in pursuit of Butch Cassidy and his Wild Bunch. It was a Pinkerton employee who unearthed a photograph the Wild Gang had taken of themselves and it was he who had it reproduced on thousands of posters in both North and South America.

With the backing of its commercial clients, Pinkerton introduced a system of substantial rewards for arrests as well as for information, resulting in a network of reward-seeking sheriffs and informants. They maintained good files and were willing to share their information with others in their own field as well as with law enforcement. The records were the closest thing to a national crime information service that existed at the time and were regarded as such even by official law enforcement.[14]

THE TWENTIETH CENTURY

Others became involved in supplying security. The railroads soon had their own in-house police as did steamship lines, freight forwarders, banks, factories, mines, and retail establishments. Major competition on a national scale came into being in 1909 with the William J. Burns International Detective Agency. Burns had received a great deal of publicity in connection with his work as an operative of the United States Secret Service in cases involving municipal corruption in San Francisco, Homestead scandals in the West, and criminal manipulation of timberlands in Oregon. This agency soon represented the American Bankers Association and the American Hotel Association.

Official law enforcement also grew during this period. By 1912, federal, state, and local law enforcement agencies in the 48 states had become established and their jurisdictions legislatively enlarged. The main efforts of law enforcement and security were now carried out by official agencies. Private security stood by to assist. The entry of the United States into World War I marked the federal takeover of the railroads and the express companies, with all in-house security staffs becoming government employees. The end of the conflict saw these properties and employees returned to private ownership and control.

The era of prosperity of the 1920s was also the same period in which an unpopular and unenforceable law, concerning prohibition, created the phenomenon of making the commission of a crime—the sale, purchase, or ownership of consumable alcohol—a social virtue. In this period, the psychological acceptance of organized gangs was a way of life for many. This acceptance of criminal behavior was in no way affected either by the depression that followed or by World War II.

In the last two decades, the modern era of private security has seen changes that include do-it-yourself home security devices on sale in department stores and electronic aids that have proliferated to the extent that last year's model is to this year's as a Model T is to a Thunderbird. People employed in private security far

---

14. Ibid., pp. 363, 383-84.

outnumber those employed in all phases of law enforcement. The need for both bodies of protectors remains obvious, even though their respective effectiveness is subject to continuing question. The history of the ancient craft of private security may be illustrative of opportunities for those of the industry with foresight.

The Development of the U.S. Security Industry

By ROBERT D. McCRIE

ABSTRACT: The security services industry has developed along three distinct lines: alarm, armored car, and security guard services. Guard services also include private investigative services. As a whole, the industry today is nonmonopolistic and competitive in meeting society's perceived need for greater protection. The industry may need greater help from law enforcement in assuring that prospective employees are fit to be hired. It may also require further legislation to ensure higher standards among the disparate companies.

Robert D. McCrie is a journalist and educator specializing in security management. He is a graduate of Ohio Wesleyan University, has a master's degree from the University of Toledo, and has conducted post-master's-degree work at universities in the United States and Europe. He is editor of Security Letter, *an industry newsletter, and the* Security Letter Source Book. *He is assistant professor of security management at John Jay College of Criminal Justice of the City University of New York.*

SECURITY and civilization are intertwined. At a time when the average American citizen is increasingly concerned with personal and public safety, a survey of the security industry in the United States may offer some insights into the changing concerns that have spurred citizens and corporations to devise more effective ways to protect lives and property. The security services industry has developed in three distinct forms: alarm monitoring and servicing; armored car services; and security guard services.

THE ROLE OF ALARMS IN SECURITY

The first modern alarm was invented in the early eighteenth century by an English promoter named Tildesley. A set of chimes was mechanically linked to the door lock. The inventor's advertisement proclaimed:

The bells associated with it are constructed in such a manner that no sooner is the skeleton key of an intruder applied to the lock than the [bells] begin to chime a plaintive air that inspires such sentiment in the minds of the housebreaker that will doubtlessly prompt him to take precipitous flight.[1]

Tildesley's chime contraption was one of several variants found in the early American colonies. A bank in Plymouth, Massachusetts, had an alarm that carried a signal by wire from the safe door to the cashier's house next door. This was the nation's first bank alarm.

In October 1852, an inventor in Somerville, a Boston suburb, filed a patent for an "improvement in electro-mechanic alarms." Augustus R. Pope used electricity to sound a continuous alarm when a door or window was opened without authorization. This system contrasted with the then-used mechanical clock that sounded alarms by the uncoiling of a spring.

Pope's system had another innovation: magnetic contacts on the doors and windows were wired in a series circuit. The magnet was attached around a U-shaped metal bar. When electrified by a battery, the bar became an electromagnet, which, in turn, was attached to a bell. The circuit was normally open when doors and windows were closed. An opened door or window closed the contact, and the circuit produced an alarm. To keep the alarm ringing, Pope placed a circuit breaker between one of the electromagnet's poles and the armature that was attached to the bell hammer. Each time the hammer rang the bell, it simultaneously opened the circuit, moving the hammer away from the bell. This closed the circuit again, continuing the alarm.[2]

It is not known if Pope ever marketed his system. If so, sales were few, because in 1857 he sold the alarm patent to an unlikely purchaser, Edwin Holmes. Eight years earlier, Holmes had started a business in Boston to sell sewing supplies. By the time he had purchased the alarm patent, Holmes's business concentrated on ladies' hoops.

Holmes apparently purchased the alarm patent as a speculative investment, but when he tried to develop the system, he ran into several problems. One of the toughest was that the alarm would not work with bare copper wire; the wire had to be insulated. Thin wires insulated with silk were just then being used by the infant telegraph industry, but Holmes needed a thicker insulated wire for com-

---

1. William Greer, *A History of Alarm Security* (Washington, DC: National Burglary & Fire Alarm Association, 1979), p. 7.

2. Ibid., p. 25.

mercial alarm applications. Holmes knew how to cover hoop wire with cotton; he covered number 18 bare copper wire by the same process. Later, he developed a wire-coating factory in his backyard.

Crime was low in Boston. Further, Boston businessmen did not trust electricity: they could not imagine that someone opening a window on the second floor of a building could trigger an alarm in an office a block away. The growth of Holmes's alarm business was disappointing, so he sought more fertile ground for his enterprise in New York City.

Once in New York, Holmes concentrated on providing alarm signals for the wealthy. Over time, a number of ingenious features were developed. The first multiplexed alarm system was introduced. The same signal circuit could identify specific windows and doors that were opened. Alarms were turned off at the office automatically when servants were expected to move about. Further, when electric lights were introduced in 1880, Holmes added a device that illuminated parts of the house when an alarm signal went off. Thus, over a century ago, many of the elements in an automated alarm system were already in place: magnetic contacts, timing mechanisms, multiplexed signals, bells, and lights.

The Holmes organization was not the only firm providing installation of alarm equipment, monitoring of alarms, and responses. By 1900, several competing companies existed in New York City alone. Some of them developed as fire signaling services in conjunction with Holmes's concentration on burglary signals.

The district telegraph companies soon produced a signal box that would accept three different messages: fire, police, and messenger. The same monitoring office could handle these multiple signals in different ways. The operator would send signals to the fire brigade, call the police, or send a messenger to the subscriber's address.

Eventually the distinction between fire-signal-only and burglary-signal-only companies largely disappeared. The development by municipalities of public fire signal services further decreased the market drive for these separate types of signal companies.

While still in Boston, Holmes met the inventor of the telephone, Alexander Graham Bell. They both used the services of the same electrical technician, Thomas A. Watson. Holmes watched the development of the telephone every day in Watson's shop and evidently was impressed. Eventually, Holmes offered his alarm stations to become the first telephone exchanges, first in Boston and later in New York.

Holmes became an investor and officer in the Bell Telephone Company of New York, an interest he sold in 1880. While no longer part owner of the burgeoning phone system, Holmes maintained contact with the companies that formed American Telephone & Telegraph (AT&T) in 1900. For example, AT&T subsidiaries willingly installed sub-voice-grade circuits for monitoring alarms side by side with its voice-quality lines.

It was no surprise when AT&T bought Holmes's company in 1905. This combination of the leading phone company and Holmes's company aided enormously in the strength and national expansion of the alarm company. Edwin Holmes's son, Edwin T. Holmes, in charge of the company after his father's death, had turned down previous acquisition offers, including one from R. C.

Clowry, president of Western Union, the nation's largest telegraph service.

Clowry had noted the rapid growth of local messenger and patrol services and decided to consolidate the activity. In 1901, he bought controlling interest in 57 of these independent firms and incorporated them in New Jersey as the American District Telegraph Company.

While the American District Telegraph Company did not purchase Holmes, the two companies entered into a restrictive agreement in September 1906 that was to reflect the special strengths of each. The agreement addressed the markets in which each would operate separately, the services each would provide, and the products each would sell. While the two major burglar alarm companies were carving up the market, the two major fire protection firms were doing the same thing. Automatic Fire Alarm of New York agreed to operate in the Northeast and exclusively in Boston, New York City, and Philadelphia. Automatic Fire Protection was to receive the remaining territory.

Within a few years, the burglar and fire alarm business had become a closed industry, with only a few companies controlling the major industrial and commercial accounts. While independents existed and even eventually formed their own trade association, the major national companies remained dominant for the next half century.[3]

The most dominant personality in the alarm industry to emerge during these years was James Douglas Fleming. Starting at a Grinnell subsidiary in 1919 as a sprinkler fitter's helper, Fleming steadily rose in Grinnell, which was then the dominant manufacturer of sprinkler equipment. He became its president in 1948. This began an amazing period of acquisition for Grinnell under Fleming's leadership. Within a few years, he had purchased Automatic Fire Alarm, Holmes, and the American District Telegraph Company.

In 1958, the Antitrust Division of the Justice Department began an investigation of the Grinnell holdings. The lead attorney on the case, Noel Storey, determined that the most significant factor in market dominance was the accredited market, that is, companies that met the requisite standards of Underwriters Laboratories. Commercial and industrial organizations had little choice about turning to sources listed by Underwriters Laboratories if they were to obtain the lowest insurance rates. Storey concluded that Grinnell controlled 90 percent of the accredited central station market. Grinnell countered that it was not a monopoly when the entire market, including all businesses and residences, was considered. The courts did not accept his argument.

In 1964, a federal judge ruled against Grinnell on every legal issue raised by the Justice Department. He further ordered the company to divest itself of all its holdings. After an appeal to the Supreme Court, the terms against Grinnell were modified somewhat, but the company was substantially liquidated. As a result, new companies—including Honeywell Protection, Wells Fargo, Westinghouse, and 3M Corporation—were able to enter the industry.

The original antitrust restraining terms against the culpable parties have been lifted. The industry is more competitive than it ever has been, helped by new low-cost technology that has produced easy-to-install sensors and modular central alarm equipment. The new options for line signal systems as a result

---

3. Ibid., p. 72.

of the breakup of AT&T have also aided in market opportunity for independent companies.

Total North American security industry revenues from the commercial, industrial, and integrated systems market segments were approximately $4 billion in 1986, with a predicted growth to $7 billion by 1990. Currently, about $500 million is spent on residential security services; the major revenues derive from commercial, industrial, and institutional services.

Over 13,000 companies provide electronic security services. Only 11 percent of these have annual revenues of over $1 million, while 49 percent have revenues under $100,000.[4] The trend strongly favors a pattern of consolidation by larger companies. Well-managed regional organizations will be able to thrive, however, by emphasizing superior service.

THE ARMORED CAR INDUSTRY

As already described, crime control was a challenge in the mid-eighteenth century. The protection and transportation of portable, valuable assets became very difficult. Many such assets—money, securities, jewelry, and the like—could be taken by one person and moved quickly. A thief or thieves could outrun the jurisdiction or the reach of law enforcement. Protection depended upon speed, secrecy, and limited physical protection—factors bound to be compromised eventually if the risk to a thief were worth the effort.

Washington Perry Brink began in Chicago in May 1859 as a parcel deliverer, with capital including one horse and a light wagon. This was the same era in which other entrepreneurs—such as Henry Wells, William G. Fargo, and Alvin Adams—saw the need for express delivery service. Wells, with others, founded the American Express Company in 1850 and, in association with Fargo, organized Wells Fargo and Company in 1852. Adams, with other associates, organized the Adams Express company in 1854. They all moved packages, money, and other valuables.[5]

Brink's business expanded, and he began to hire workers. One innovation was that he hired only unmarried, husky men of demonstrated probity. The significance of being single was that the workers would live in the company boarding house, Brink's home. This enabled Brink to reclaim some wages for boarding expenses, to keep tabs on the workers, and to make sure that each worker finished the day's deliveries.

Over the next generation, the company grew as a package express firm. In 1881, the first payroll for Western Electric Company was delivered. The first recorded delivery of bank funds occurred in 1900 with the delivery of sacks of silver dollars. Gradually, the pickup and delivery of payrolls became a regular part of Brink's business.

By 1913, Perry Brink had retired, and the company's revenues were then about equally divided between package express services and money shipments. Subsequently, the company concentrated more and more on money deliveries. The company acquired branches in other cities and negotiated area contracts with banks, retail chains, and other businesses.

During the first 26 years of money-

---

4. ADT, Inc., *Annual Report 1986* (Parsippany, NJ: ADT, 1987), p. 4.

5. R. A. Seng and J. V. Gilmour, *Brink's the Money Movers* (Chicago: Lakeside Press, 1959), p. 20.

moving operations, no Brink's employee was attacked. The first instance of attack occurred on 28 August 1917, when four bandits surprised Brink's deliverers and escaped with a factory payroll. Three of the bandits were later arrested, and much of the money was recovered. Brink's quickly moved to improve security, and a new era began that required money-moving firms to be more security conscious. This development, over time, tended to restrict money-moving services to companies that had the secured facilities, training, and insurance to handle them more safely.

Only a few armored car companies were able to handle such movements of money on a large scale. Brink's received its first regular contract from the Federal Reserve Bank of Cleveland in 1949. A decade later, the business had expanded enormously, with Brink's as the dominant company in the field. Wells Fargo was second largest. By 1970, two companies, Wells Fargo and Purolator Armored, had operations in several states and small companies served local regions.

By the early 1960s, Brink's had become a monopolistic company, controlling about 50 percent of the entire armored car market. Together with the next two firms, Wells Fargo and Purolator Armored, it controlled about two-thirds of all armored car activity in the United States, with similarly strong market positions in Canada.

With respect to strong government regulation, Brink's was more the beneficiary than the victim. As a contract motor carrier, the company was subject to the regulation of the Interstate Commerce Commission. The commission was permitted to regulate the industry in many ways. This included designation of what could or could not be carried, accounting methods, reorganizations, and, most important, the issuance of interstate operating rights. Armored car companies were required to file their minimum rates but not their actual rates or contracts.

The insurability factor aided the growth of Brink's. By 1970, Brink's was able to claim it carried up to $50 million of insurance coverage for loss in any one vehicle or other conveyance.[6] This was more than any other armored car company, and it gave the firm a substantial advantage over local independent firms.

For example, the manager for a Federal Reserve Bank branch might limit the right to transport assets to member banks to a company that had the ability to insure the cargo for a minimum amount. Therefore, Brink's and a very few others would often be the only firms to have the requisite insurance capacity to deal with the Federal Reserve Banks. The result was decreased competition and guaranteed pricing levels.

In the 1960s, the Antitrust Division of the Justice Department began a protracted series of actions aimed principally at Brink's and Wells Fargo. Their competitors and customers then instituted class action suits for damages and recoveries. The last class action and antitrust suit was settled in 1978.

Today, the armored car industry may be described as a crypto-oligarchy but not as a monopoly. Trucking deregulation in 1980 and changes in Federal Reserve Bank policies opened opportunities for greater competition. A host of new armored car companies emerged in the early 1980s, but some of them were not sufficiently experienced or capitalized to survive.

In addition to the top three com-

6. Offering circular for Brink's Incorporated common stock, issued by Blyth & Co., Inc., 1970, p. 7.

panies—Brink's; Loomis Armored, which absorbed Purolator Armored; and Wells Fargo—perhaps 125 other armored car companies exist. Some are thriving interstate businesses. Most, however, have fewer than a dozen trucks, operating their money shipments as a subordinate activity to other business.

THE GUARD INDUSTRY

The guarding of people, property, and things began long ago. The pioneer in guard organizations in the United States, however, was Allan Pinkerton. The son of a policeman, Pinkerton was born in Glasgow, Scotland, in 1819. He was politically disillusioned as a youth and became a member of the Chartist movement, which sought, in part, wider voting rights and greater representation in Parliament.

Chartism reflected a basic schism between the Scots and the English establishment. Harassment of Chartists grew, so Pinkerton and his young bride emigrated to the United States in 1842. The couple eventually reached Chicago. Pinkerton became a cooper for a local brewery. A year later, he left and moved to Dundee, a nearby community, opening a cooperage. Business flourished.

In 1847, Pinkerton was walking in the woods, looking for trees to harvest for the cooperage, when he discovered a suspicious gang that was meeting in a clearing. They were dealing in green goods—that is, counterfeit money—then a serious problem in Illinois. After discussing his findings with local officials, Pinkerton contrived to buy some counterfeit money. This act led to the arrest of the counterfeiters.

Pinkerton became a local hero and an expert on catching counterfeiters. His services were requested and given to Kane and Cook Counties. In 1851 and 1853, he was hired to investigate counterfeiting.[7] In 1852, he traced and rescued a kidnap victim at the request of the Cook County sheriff. Over time, Pinkerton became more interested in investigations than in barrel making.

In 1855, Pinkerton opened his investigation agency. His first clients included the post office, for which he was a special agent, and various railroad companies.

The railroads had serious security problems. Once a train left a city, it was highly vulnerable. The railroad could expect little help from local police if a train were attacked by bandits in open areas. Railroad property was also frequently vandalized, often by people objecting to the encroachment of trains on land they considered theirs. Rural areas were especially dangerous because, at the time, no state had organized police forces.

From a business standpoint, railroads made excellent clients; they had a need that Pinkerton could fulfill. The railroads urgently required security and investigative services. There were, furthermore, virtually no legal impediments to the types of services Pinkerton decided to perform—in fact, he was authorized as an agent of government. Finally, the railroads would pay enough to make the venture attractive.

In 1855, six Midwestern railroads gave Pinkerton $10,000 to establish the North West Police Agency.[8] Thus began a long association between the Pinkerton organization and the rail transportation industry, with Pinkerton providing guard

7. Frank Morn, *The Eye That Never Sleeps* (Bloomington: Indiana University Press, 1982), p. 54.

8. James D. Horan, *The Pinkertons* (New York: Bonanza Books, 1962), p. 31.

and patrol services. The investigative exploits of the fledgling organization attracted extraordinary attention.

As an industry pioneer, Pinkerton was innovative by necessity. He provided the first modern executive protection service, serving Abraham Lincoln at different times. The firm provided intelligence and counterintelligence services to the North during the Civil War. Pinkerton employed the first woman investigator, Kate Warne, who now lies buried with him. The Pinkertons developed the first independent crime laboratory and used analytical methods in their cases.

Pinkerton soon learned that often the most devastating thief was an employee. He uncovered cheating postal clerks, thieving bank managers, dishonest railroad employees, and many others. "The eye that never sleeps" became the Pinkerton slogan, and Pinkerton began undercover services to spot dishonest employees. Increasingly, in the period 1870-90, the eye was also watching strikers.

As the founder's son, Robert Pinkerton would say, "We never looked for any strike work; it was something which has grown about our shoulders." The Pinkerton agents never claimed to be strikebreakers. Rather, they thought of themselves as private citizens—though usually sworn as temporary deputies by local sheriffs—protecting private property. Still, the business grew and the Pinkertons became a familiar presence at labor disputes throughout industrial America. In twenty years, Pinkerton's became involved in seventy strikes.[9] The knights of labor faced the knight of capitalism.

The watershed of Pinkerton's labor activities came during the company's strike services for the Carnegie, Phipps Steel Company, managed by Henry Clay Frick. Frick feared a strike at the Homestead, Pennsylvania, plant and bypassed local law enforcement to obtain security. He hired 376 Pinkerton guards and moved them secretly into the strike site, where a sit-in was taking place.

On 11 July 1892, as the barges carrying the private police moved near the steel company's siding, strikers fired on the crafts. A 12-hour siege resulted, and when it concluded, three guards and 10 workers were dead. Revulsion against "Pinkertonism" flared. The Pinkerton company and other strike services became the subject of a congressional hearing. Despite the fact that reaction against labor unionism at the time was intense, the Pinkerton reputation was harmed by the Homestead incident, and the image-conscious organization curtailed its strike services.

The turn of the century was a time of heroic, larger-than-life characters. Such crime fighters often caught the attention of the public. In the early years of the century, New York and Chicago had hundreds of detectives searching for glory and rewards. Many of them were solo practitioners. An increasing number, however, were being employed by government. One of these was William J. Burns, who began his career as an independent detective. Burns joined the Secret Service as an assistant operative and was involved in cases that became national news events. He served President Theodore Roosevelt as a celebrated investigator and graft fighter. Burns founded the William J. Burns International Detective Agency, today Burns International Security Services. He ended his public career by returning to government and heading the Federal Bureau of Investigation, preceding J.

---

9. Morn, *Eye That Never Sleeps*, p. 106.

Edgar Hoover.[10]

By the decade 1960-70, the nation had thousands of investigators and private security guard companies. The industry had grown without much critical analysis. Further, it existed without much regard for its impact on public policy.

With a grant from the Law Enforcement Assistance Administration, the Rand Corporation began in 1970 a 16-month investigation of private police in the United States. The authors, James S. Kakalik and Sorrel Wildhorn, were lawyers. Many individuals from different areas of public and private life contributed to the findings in this first analytical look at the industry.[11]

The Rand Report, as it is known, gave dour pronouncements about the security industry and its catalog of problems. For example, the guard industry was found to lack adequate training for guards. Persons hired in the industry had limited education. Preemployment screening was weak, and there was a lack of meaningful licensing standards. In the investigative industry, there was disregard for rights of privacy during the collection of information, and operatives had low educational and ethical standards. In the alarm industry, there was a high frequency of false alarms, with subsequent waste of law enforcement resources, and standards for employees were uncertain.

Over the previous generation, public law enforcement had made an extraordinary leap in improved training and selection. This served to accentuate the differences between the increasingly professional law enforcement officer and the often minimally paid, untrained guard.

Solid reasons existed, however, for law enforcement and private security to understand each other and search for areas of cooperation. The frequently superior attitude of some public law enforcement officials toward square badges, or private guards, was not productive.[12] Law enforcement and private security are fundamentally different; they are supposed to be. Yet a positive, supportive relationship could aid both groups.

The Rand Report described an untenable situation—a burgeoning industry affecting public life and liberty but with few real standards. The rather diffuse and legally oriented concerns of *The Rand Report* shocked much of the public. In 1972, the Law Enforcement Assistance Administration founded the National Advisory Committee on Criminal Justice Standards and Goals. One task force of the committee studied private security. In its report, the task force recommended almost eighty goals and standards for private security. They covered such areas as licensing, regulation, consumer services, personnel training, crime prevention systems, and conduct and ethics.[13]

In the dozen years since the report was issued, many factors have driven the industry to greater professionalism. These factors include

—self-interest; the industry would rather see legislation of its own

---

10. Gene Caesar, *The Incredible Detective* (Englewood Cliffs, NJ: Prentice-Hall, 1968), p. 18.

11. James S. Kakalik and Sorrel Wildhorn, *Private Police in the United States: Findings and Recommendations* (Washington, DC: Government Printing Office, 1972).

12. *McKinney's General Business Law*, articles 7 and 8.

13. National Advisory Committee on Criminal Justice Standards and Goals, *Private Security: Report of the Task Force on Private Security* (Washington, DC: Government Printing Office, 1976).

design than that imposed by outsiders—as a result, some activism in passing regulations is currently taking place;
— litigation; successful actions by plaintiffs against security companies for negligence and other charges are forcing the industry to improve its standards;
— insurance, in that insurance companies may force substantial management and operational changes on inefficient security companies; one change, certainly, is that guards today are far less likely to be armed than in the past—insurers have insisted on this change;
— customer requirements; government agencies that contract with private security services often have demanding standards for selection, training, continuing education, supervision, support equipment, and other matters—this has been having a salutary effect on the industry as a whole; and
— the law; laws in banking, the energy industry, and other employment sectors set standards for security that will upgrade standards in non-regulated industries.

The number of people employed in the security guard and investigative industry is large. According to a comprehensive study of the industry, private sector security employment was recently estimated at 1.1 million.[14] This exceeded the employment in local, state, and federal law enforcement of roughly 650,000.

The contract portion of these services generates $5 billion in expenditures. The industry is highly competitive, with no single company controlling as much as 10 percent of the market. The top thirty firms represent about 50 percent of the entire market.[15] About 13,500 additional firms provide services nationwide. They range greatly in terms of quality, experience, insurance strength, and depth of management. Additionally, billions of dollars are allocated for internal, or proprietary, security services. In a few states, security guards must meet standards similar to those required of contract guards, but this is exceptional at present.

The contract security guard industry continues its steady growth. The number of private police has exceeded the number of public law enforcement personnel for some years now, and the disparity is growing.

PUBLIC POLICY QUESTIONS

Serious public policy questions exist for the security industry.

In our quest for greater security, are we in danger of placing some precious civil rights in jeopardy? Does a prospective employer have an absolute right to submit employees to polygraph and drug testing? Does he or she have the right to police and Federal Bureau of Investigation records? Are closed-circuit television cameras an invasion of privacy?

These and other questions are under rigorous scrutiny and debate in state and federal legislative committees, and no simple answers seem to be forthcoming soon. It seems that the classic debate between the relative merits of freedom and order, voiced long ago in Greece, are with us still and show no signs of disappearing.

14. William C. Cunningham and Todd H. Taylor, *Private Security and Police in America* (Portland, OR: Chancellor Press, 1985), p. 113.

15. Robert D. McCrie, ed., *Security Letter Source Book, 1987-1988* (Stoneham, MA: Butterworths, 1987), p. 141.

True security for a society, of course, lies only in the free and responsible submission of the citizen to fair and reasonable laws. To paraphrase James Madison, if men were angels, we would need no locks or guards or security companies. Until that day appears, we will be stuck with all three.

Personnel Selection in Private Industry: The Role of Security

By ROBERT W. OVERMAN

ABSTRACT: Personnel selection procedures used in private business and industry suffer from a lack of attention to security considerations. This neglect is a result of training, philosophy, structural rigidity, turf protection, and political and legal trends aimed at protecting individual rights. Past performance remains the best predictor of future performance, but discovering a job applicant's history is becoming increasingly difficult. Legal rulings have made criminal, medical, and financial records safe from inspection by personnel managers. Even previous job performance records are being made off-limits by employers who have come to fear lawsuits filed by previous employees on grounds of defamation. The personnel selection techniques used by private security companies would greatly aid private business and industry in attempts to acquire a quality work force. The possibility of genetic screening poses new challenges to our moral, ethical, and legal systems of thought.

Robert W. Overman is vice president, manager, human resources for Guardsmark, Inc. He is a licensed psychologist with experience in personnel administration, psychological testing, management consulting, and law enforcement. He received a Ph.D. in industrial-organization psychology from the University of Tennessee, and he is a member of the American Psychological Association.

THE importance of security and loss prevention in the personnel selection process in private business and industry does not receive the attention that it is due. For the most part, security and personnel functions are kept separate and distinct, with roles and concerns that do not intersect. The perspectives and philosophies of these two key corporate areas are antithetical, and the rich potential synthesis often goes unrealized.

The security perspective offers immense value in furthering corporate goals through personnel selection. Hiring the most productive and efficient employees can be an impossible task if the job applicants are not appropriately screened from the standpoint of security.

Thorough screening is absolutely essential. More often than might be expected, job applicants falsify their credentials, make unwarranted claims regarding their achievements, hide past misconduct or problems behind the privacy of their criminal and medical records, and rely on previous employers not to discuss their job performance.

DIFFICULTIES OF PERSONNEL SELECTION

Personnel departments today are operating at a tremendous disadvantage. At a time when everyone in our society has a paper trail from cradle to grave, when records are computerized, centralized, and accessible through advanced technology, when tests are standardized and comparable on a national scale, personnel managers still face enormous obstacles in deciding with respect to any particular applicant, "Who is this person?"

An applicant could be a convicted felon, but the personnel manager has no way of discovering that fact because he or she has no access to criminal records. The applicant could be a diagnosed paranoid schizophrenic, but medical records are private. A high school diploma is no longer even a guarantee of literacy. Dismissal from a previous job for stealing can be covered by the employer's reticence stemming from fear of a defamation suit. Claims made by an applicant often cannot be verified by a polygraph examination.[1] The personnel manager is left trying to evaluate an applicant's character, veracity, and performance potential by means of a personal interview, a highly subjective, non-scientific method of determining how an applicant would fit into the corporate culture.

The obvious response on the part of business and industry is to avoid the unknown. This is undoubtedly one of the reasons that most job openings are never advertised but are filled through private networks. An unfortunate side-effect of the movement to protect privacy has been the creation and preservation of this private job placement system. Those who most suffer from the way the system operates are precisely those who would most profit from free access to employment opportunities: the disadvantaged, the educated poor, and all those who lack influential business and family connections.

IMPORTANCE OF SECURITY

The issue of privacy is a central one for citizens in a democratic society, but it is a particular problem for the personnel department of a private company. An applicant's right to protection from invasive probing under an aggressive

---

1. The District of Columbia and 12 states prohibit the use of the polygraph in personnel selection. As of this writing, federal legislation is pending that would ban the use of the polygraph in

personnel policy must be balanced by a company's right to know whom it is hiring. A dishonest person, a drug user, or a pyromaniac can wreak havoc on or destroy a company, costing innocent people their jobs or even their lives.

In monetary terms, it is difficult to estimate the annual cost of crime against business. The highly regarded *Hallcrest Report*, a national study of the private security industry, reviewed the available research on crime against business and placed the annual cost at anywhere from $67 billion to $300 billion.[2] A substantial portion of that cost can be attributed to employees. Employee crimes that contribute to the total include fraud, embezzlement, retail shrinkage, sabotage of equipment and other resources, and the selling of insider information such as financial data, codes, and technical processes.

In addition, courts have held companies liable for the acts of their employees that result in loss or damage to others. A finding of liability can also be based on a judgment that an employer should have foreseen that a particular employee might commit a certain kind of act. For example, the owners of an apartment complex in Minnesota were held financially liable for a rape committed by the apartment manager. The owners had not investigated a five-year gap in the man's employment history. Because he had spent those years serving a sentence for armed robbery, the court held that his committing another violent crime was "foreseeable."[3] As this case

---

the private sector except for specific polygraph exams.

2. William C. Cunningham and Todd H. Taylor, *Private Security and Police in America (The Hallcrest Report)*(Portland, OR: Chancellor Press, 1985), p. 25.

3. Larry Reibstein, "Firms Face Lawsuits for Hiring People Who then Commit Crime," *Wall Street Journal*, 30 Apr. 1987.

illustrates, in our litigious modern society, the hiring decision can have enormous consequences.

PERSONNEL SELECTION VERSUS SECURITY

Despite the obvious advantages that can follow from including security considerations in the hiring process, it is the rare company that does so in any consistent manner. In many instances, the personnel department is in charge of nothing so much as the flow of paperwork: placing ads for positions, seeing that insurance forms are properly filled out, issuing parking decals for the company lot, and so on. The manager of the department that has the open position makes all the important decisions, down to which applicants deserve a personal interview.

In other companies, the personnel department maintains tight control over the entire process. While the final decision rests with the manager of the relevant area, the choice is limited to those candidates who pass the screening procedures of the personnel department.

In both instances, the impulse is to fill the position as quickly as possible with the best person available. Often there is only one real candidate, who has been recommended by the president or someone else in senior management. In such a case, the personnel department acts as a rubber stamp. Many times, because special projects are lagging or because an increase in the regular work load is adversely affecting staff morale, the department manager with the opening may bring particular pressure on the personnel department to fill the position immediately.

The dynamics of the hiring process militate against any procedure—such as security screening—that might create a delay. The risks of a delay are that the hiring company might lose the favored candidate. The candidate might accept another offer, take a jaundiced view of the company, or even receive a promotion from his or her current employer.

A second and greater difficulty in promoting the use of a security perspective in personnel departments is a simple lack of understanding. Personnel managers are highly educated in all areas of personnel administration, but their knowledge of and experience with security are sadly deficient. Security is seen as an entirely separate function that has nothing to do with the personnel area. The importance of screening job applicants is well understood, but the need to do so from a security perspective is almost completely ignored.

The last reason for resistance on the part of personnel administrators is largely one of turf protection. Defining security and personnel as distinct functions prevents the sharing of authority and responsibility for personnel policies and procedures.

INCORPORATING SECURITY CONSIDERATIONS

Artificial distinctions between the two areas can be eliminated through a sincere effort on the part of management. The easiest way is to combine the two into one. The results can be dramatic. A department store in Alabama realized that much of its shortage could be traced to its hiring decisions. Combining the security and personnel departments and using new security devices cut the store's shrinkage—losses from theft and bookkeeping errors—from 2.5 percent of sales to 1.6 percent, a sizable savings.[4]

> 4. "The Boom in Digging into a Job Applicant's Past," *Business Week*, 11 June 1984, p. 68E.

The security perspective includes a view of human nature that is less rosy than that which characterizes personnel administration. Security accepts no applicant's claims on their face value. Whereas personnel professionals would tend to believe such claims to a greater or lesser extent, a security professional would give them no credence at all until they were supported by evidence. That does not necessarily make security adversarial. The role of security is more that of a doubting Thomas, who demands proof, than of a devil's advocate, who tries to undermine the proof.

Background investigation

Introducing basic security techniques would greatly enhance the average personnel selection process. A background investigation is essential, with its extent scaled to the importance of the position. Some pieces of the investigation are easily carried out. For example, to verify that an applicant actually earned the degree he or she claims may require only a phone call to the school or college, but it is doubtful if even half the companies in the nation bother. Military service records can also be checked without difficulty.

Criminal records

Criminal records of applicants cannot be checked by the vast majority of private employers. State statutes vary widely in requiring or permitting companies in specific fields to run a check on the criminal records of applicants. Most companies have no access to criminal record histories.

With the polygraph under attack, in many cases, the only way for private businesses to ensure that the applicant has a clean record is to require a detailed explanation, with supporting documentation, of any gaps in the employment record.

Employment history

Problems arise with the questions of employment records. Increasingly, companies are reluctant to disclose negative information about previous employees for fear of defamation suits. To avoid the possibility of being sued, these companies verify only the dates of employment and the titles of positions held. Employees with good records are penalized by this silence, but companies feel, perhaps rightly, that it is not their place to help former employees advance their careers.

An employer's refusal to discuss the work habits and job performance of a former employee leaves the hiring personnel manager with the important questions unanswered. The job applicant might have been dismissed from a previous job for any number of reasons that make him or her a risk, even a dangerous risk. It may be in the former employer's interest to leave key questions unanswered, but the hiring company knows that the best predictor of future performance is past performance. The applicant's past performance must be unearthed.

Digging up such information involves going beyond the personnel department to former supervisors and coworkers. Personal acquaintances of the applicant are likely to be much more open than some anonymous personnel assistant who is trained to reject all probing. It is also the case that people who worked with the applicant on a daily basis have a richer understanding of his or her character and work habits than can be revealed by any standardized performance-rating system. The best source is a former supervisor who is no longer employed by the applicant's previous employer. Having left the company, the former supervisor is more likely to be open and candid about the applicant.

Eliciting a useful response from a former supervisor or coworker is a matter of applying good interviewing technique. Here, too, the security experience can strengthen the hand of the personnel department. Security investigators are professionals at probing beneath the surface to get at the truth and are not easily sent away with simple responses of yes or no. Questions can be designed to prevent an applicant from snapping off answers that essentially reveal nothing.

Investigators are also attuned to sensing the areas that require further probing. Responses that are intended to harm the applicant's chances of being hired are also obvious to the skilled investigator. Such responses can be weighed against the other evidence gathered in the investigation to see if they have a solid basis in fact or are motivated by malevolence toward the applicant.

The same pointed, probing search for truth can be applied to personal interviews with the applicant. Most interviews barely scratch the surface and are designed only to help the personnel interviewer gain some sense of the applicant's personality. Security interviews, however, aim at uncovering facets of character in addition to assessing personality. Asking the detailed, hard questions about performance claims, discrepancies in statements, gaps in the employment record, and so on can lead the interviewer in unexpected directions that a cursory interview would never point out.

Conscientious, security-minded management will take all reasonable measures to ensure that the company, its assets, and its employees are well protected. That private business and industry are increasingly interested in employing security measures is illustrated by both the rapid growth of the private security field and the swift adoption of drug testing as a preemployment screening device.

Minnesota Multiphasic Personality Inventory

Another measure used in private security and law enforcement, but not widely employed in general industry, is the Minnesota Multiphasic Personality Inventory (MMPI).[5] The MMPI is a personality test that consists of 566 questions designed to measure such emotional and behavioral problems as depression, insecurity, aggressiveness, anger, and paranoia. Originally developed as an aid in diagnosing clinical patients, the MMPI has been shown to be useful outside the clinical setting. Even there, however, the results must be analyzed by a licensed psychologist, although the test need not be administered by one.

The test can be particularly valuable in assessing the ability of individuals to operate in high-stress occupations. It does not predict—and was not designed to predict—employee performance, and so it cannot be used as a tool for preemployment screening. The MMPI can, however, help ensure that new hires are not placed in positions with responsibilities for which they are psychologically unsuited.

A couple of caveats are in order. First, although the MMPI is useful in indicating potential mismatches of employee and position, the test is not effective in creating a match between a particular position and a type of personality.

The test also has inherent limitations that must be factored into its cost-effectiveness. Because it is designed to identify aberrant personality types, the great majority—approximately 85 percent—of persons tested show results in the normal range.[6] That result is to be expected, but it increases significantly the cost of each identified case of potential psychopathology. As in other areas, the costs must be weighed against the expected benefits. More and more companies are seeing the value in such exclusionary screening, with drug testing a notable example. The costs of administering the MMPI are minimal compared to the potential liability that could result from product tampering, an industrial disaster, or a crazed employee's shooting spree.

THE FUTURE OF SCREENING

Employee screening has grown more difficult since the court rulings on privacy of the 1970s. The current trend is mixed, with the sudden spread of drug testing representing great strides for security. On the other hand, the movement to ban use of the polygraph in preemployment screening threatens to eliminate an instrument that has proven very effective in the hands of professional examiners.

Restrictions on access to various kinds

---

5. Guardsmark, Inc., administers the MMPI to every new employee. Over the years, the test has been taken by approximately 75,000 employees, forming perhaps the largest collection of MMPI results in existence.

6. James N. Butcher, "Use of the MMPI in Personnel Selection," in *New Developments in the Use of the MMPI*, ed. James N. Butcher (Minneapolis: University of Minnesota Press, 1979), pp. 165-201.

of information—criminal, medical, financial, and employment—have forced security-conscious personnel managers to use screening methods, such as in-depth background investigations, that they might not otherwise think necessary.

Personnel managers who are concerned about security naturally use the best cost-effective means available to ensure a quality work force. No screening method is devoid of social implications, and technologies are being developed that will present new challenges to our moral, ethical, and legal systems. The greatest of these challenges will be posed by the revolution now occurring in our understanding of human genetics.

Genetic screening

The advances being made in genetics hold vast promise for human welfare. Completing a map of the tens of thousands of human genes no longer seems impossible.[7] With such a map in their possession, scientists could gain new insight into the relationship between heredity and disease. Tests already exist for many diseases, such as polycystic kidney disease, emphysema, sickle-cell anemia, colon cancer, Duchenne muscular dystrophy, cystic fibrosis, Huntington's disease, and childhood eye cancer.

In the near future, tests may well be developed for hypertension, dyslexia, hardening of the arteries, cancer, manic-depression, schizophrenia, juvenile diabetes, familial Alzheimer's, multiple sclerosis, and muscular dystrophy.

The benefits of being able to identify and locate a defective gene that is linked to a disease are obvious. Some researchers foresee the development of individual genetic profiles by the year 2000. Some persons might find, through these profiles, that smoking or cholesterol pose no problem for them.

Others might discover that they are likely candidates for an early heart attack. Those whose genetic profiles reveal a predisposition to a particular disease could seek early treatment or undertake a preventive program.

Genetic profiles would also be valuable in reducing potential occupational hazards, such as those involved in working with certain chemicals or other substances that are connected, however remotely, to the development of certain diseases. The personnel manager's current hunch that an applicant is unsuited for a particular position would in some cases become a scientific certainty.

The legal and social implications of such possibilities cannot even be cataloged, although many can be imagined. Insurance companies would certainly have an interest in gaining access to genetic profiles, as would personnel managers. Few companies would knowingly hire applicants who were likely to develop disabling diseases.

The creation of genetic profiles would produce thorny ethical quandaries. Genetic screening by employers could lead to a new caste of economic untouchables: persons who, however well educated and work oriented, would not be able to find employment in the private sector. On the other hand, an employer's ignorance of an applicant's genetic profile might allow that person to be placed in precisely the kind of work environment that would trigger a malady.

Some cases would require special considerations. For example, a strong argument could be made that airlines should have the right to know if their pilots have a predisposition to manic-

---

7. Harold M. Schmeck, Jr., "Potent Tool Fashioned to Probe Inherited Ills," *New York Times*, 11 Aug. 1987.

depression or schizophrenia, if only for purposes of periodic monitoring.

Resolving such cases of conflicting interests will not be easy. But when the individual's right to privacy brings with it potential, incalculable harm to the public welfare, compromises will be necessary. The revolution in genetics will necessitate a new understanding of moral and ethical behavior along with changes in the legal structure to incorporate the new view of men and women in society.

It is understandable that private companies will adopt whatever screening measures are available to ensure the quality of their work force. It is equally understandable that individuals do not want to be screened. This tug of competing interests creates a healthy dynamic, helping to place individuals in the most suitable positions. That dynamic is eroding, however, as companies see one screening measure after another made unavailable. Drug testing is the only exception to this trend.

Denying companies valid means of screening applicants will lead many to use invalid means in hopes of screening out at least some of the unsuitable applicants. The theory in operation here is that something is better than nothing. The irony is that society is harmed, companies are harmed, and untold numbers of individuals are denied employment for no good reason.

Biological screening

The dire effects of access to only limited screening are illustrated by the argument for biological screening. The basis of the argument is that persons can and should be screened out using criteria over which those persons have no personal control, such as age, sex, criminal record of parents, and so on.

On the face of it, the argument is preposterous, discriminatory, and inimical to the ethos of a society that believes individuals should be judged on their own merits.

Advocacy of biological screening is based on findings that correlations exist between factors such as age and gender and higher-than-average crime rates. For example, studies have shown that crime rates are higher among males than females and that youths from 15 to 19 years of age are about 16 times as likely to be arrested for property crimes as persons aged 50 to 54.[8] Biological screening simply excludes applicants who fall into the higher-than-average crime-rate categories.[9]

There can be little doubt that some companies try to protect themselves by using the dubious and socially destructive methods of biological screening. And it may be that awareness of such methods underlies to some extent the trend toward hiring retired workers. Personnel managers who rely on biological screening are in effect throwing up their hands with regard to personnel selection.

That is a reckless response to the situation. Quality control in the selection process in private business and industry has become increasingly difficult to maintain, but the solution lies in the innovative adaptation of responsible selection procedures that are used effectively by reputable private security companies.

Current business trends portend the ever-growing importance of personnel

---

8. James Q. Wilson and Richard J. Herrnstein, *Crime and Human Nature* (New York: Simon & Schuster, 1986), pp. 104-47.

9. For advocacy of this position, see Saul D. Astor, "Biological Welfare," *Security Management*, May 1986, pp. 67-69.

selection. As corporations gear themselves for a worldwide competitive market, they are coming to realize that their most important resource is their work force. At the same time, competitive pressures are inducing cost-cutting measures to eliminate corporate fat: more and better work must be done by fewer and fewer people. Successful personnel selection will be a crucial element in the new competitive environment.

The costs of selecting and training employees are enormous. According to one estimate, an hourly worker who quits or is fired after a few months on the job can cost the company $5,000 in lost productivity and training. For a manager, the cost can reach $75,000.[10] The new emphasis on personnel selection and retention and the costs of mistakes in hiring would seem to indicate that the additional costs of security screening would be seen as minimal.

Done responsibly, security screening protects the rights of individual applicants as well as the rights of the company and the public in general, helping to ensure that positions of trust are held by those who bring no threat of risk to their companies, the public, or themselves.

10. Brian Dumaine, "The New Art of Hiring Smart," *Fortune*, 17 Aug. 1987, pp. 78-81.

Drug Testing in the Workplace

By PETER B. BENSINGER

ABSTRACT: The increasingly serious problem of substance abuse in the workplace is reviewed. There are new initiatives to deal with this major public health problem in the United States. The various drug-testing methodologies and procedures are outlined, and the value, accuracy, and impact of drug testing are discussed. The development of drug-testing programs sensitive to the medical, legal, and work-related issues raised by substance abuse in the workplace are reviewed. The concerns of industry regarding alcohol and drug abuse in the workplace relate to health and safety, accidents, absenteeism, and medical and insurance costs. Employers have rights and responsibilities in ensuring a safe work environment. Specific suggestions for employers concerned with the drug problem in industry are discussed.

Peter B. Bensinger is president of Bensinger, DuPont & Associates, a professional consulting firm providing services to private industry, national and community organizations, government, and professional sports on drug and alcohol abuse policy and on drug testing. Prior to forming this consulting service with Dr. Robert DuPont, Bensinger served as administrator of the U.S. Drug Enforcement Administration. His articles on drug abuse have appeared in Newsweek, U.S. News and World Report, *the* New York Times, *the* Washington Post, *the* Harvard Business Review, USA Today, *and other national and international publications.*

DRUG testing is not a magic wand that will guarantee a drug-free workplace, but it is an important tool aimed at that goal and being implemented by employers throughout the United States. In today's environment, the availability, use, and abuse of drugs in American society is pervasive and represents a particular drain on American industry. The costs are staggering in terms of human lives, accidents, injuries, absenteeism, productivity, and other tangible and intangible factors. In 1987 in America, there were over 18 million individuals using marijuana at least once a month; over 6 million cocaine users; 10 million individuals abusing pills, taking controlled substances without a legitimate physician's prescription relating to a specific illness; in excess of 600,000 heroin users; and from 9 to 12 million alcoholics. The cost to industry of drug and alcohol abuse has been estimated to be more than $100 billion.

Is drug testing legal? Drug testing has been found to be legal if appropriate standards, safeguards, and procedures are used. The constitutional right of freedom from unreasonable search applies only to actions by government and not to those by private employers. Regardless of whether it occurs in private industry or in government, however, drug testing should be reasonable and based on specific goals and objectives.

Employers have the right to test for drugs because they have an overriding obligation to provide safe working conditions. Employers that negligently hire employees or permit them to work while under the influence of drugs or alcohol have been found guilty in court and have been successfully sued for damages by the families of victims injured or killed by employees high or drunk on the job. The court and arbitration rulings on drug testing are still evolving.

Employees do not want coworkers snorting cocaine, stoned on pot, or under the influence of alcohol. In fact, a recent employee attitude survey by Sirota and Alper Associates, a New York opinion polling firm, found that the majority of employees questioned were in favor of drug testing by a seven to three ratio. Moreover, in this survey of over 2000 blue- and white-collar workers, hourly workers and union members favored drug testing more often than management did.[1]

Testing for drugs is one way employers can reinforce their commitment to a drug-free workplace and their commitment to the safety, health, and protection of their most important asset, their people. Of the *Fortune* 500 companies, 40 percent are now implementing drug-testing programs. Employers need to decide under what circumstances testing will take place.

WHY AND WHEN TO TEST

Employers have considered a number of circumstances under which drug testing would be appropriate. One is in the process of applying for employment. To ensure that drug users do not enter the work force, many companies are now testing individuals applying for positions.

Once hired, an employee might prove to be a candidate for fitness-for-duty testing at some point. This is being implemented in situations in which employees appear to be unsafe on the job. Indicators calling for testing are work habits that do not appear to be in accordance with normal job requirements, operating procedures, or the in-

---

1. Sirota and Alper Associates, New York City, 1986.

dividual employee's traditional work level or behavior.

Reported or observed violations of company policy sometimes warrant testing. If there is a report of observed drug use from a creditable source, the suspected individual may be required to take a drug test even though he or she may not appear to be physically impaired or unsafe. If there is a reasonable suspicion that the employee has taken drugs in violation of company policy, the company should intervene, investigate, and test to determine if drugs are present in his or her system.

Testing may be implemented after an injury or an accident. Postaccident and postinjury testing is being used particularly in safety-sensitive industries. For example, it is now required by the Federal Railway Administration for all serious train accidents and certain other safety violations. Companies have provided for postaccident and postinjury testing in the event that outside medical care or significant property damage occurs.

Perhaps one of the most valuable utilizations of drug testing occurs after treatment for drug abuse. It provides an incentive for the drug user to remain drug-free and provides protection for the employer and other employees and the general public. It provides one indication that the individual user has followed, or is following, a prescribed treatment program.

Testing in conjunction with annual physicals or special medical examinations is now being implemented particularly at refineries, at airlines, and in job situations requiring medical clearance. Also, companies are providing drug tests to employees returning from furloughs or layoffs, although not necessarily maternity leaves. An absence of 60 to 90 days can trigger a drug-test requirement for companies utilizing a return-to-work testing program.

Finally, there is random testing for drugs. This is being used increasingly in industry in safety-sensitive situations, almost exclusively among workers in positions that place the public at risk: nuclear power plant employees, military personnel, airline pilots, and law enforcement and security officials. Random testing is being implemented by 8 percent of the American companies that are using testing programs; it will provide the most effective deterrent to drug use on and off the job by the work force. The courts will balance the employee's expectation of privacy with the safety and security needs of the work assignment. In certain cases, courts have justified unannounced random testing as being in the public interest. Some arbitrators and courts, however, have been reluctant to sustain random testing among private and public employers where the public is not at significant risk and where there are not compelling reasons to justify increased surveillance or safety standards. Relevant case law is still evolving and will be subject to different interpretations until the Supreme Court provides clear guidelines.

WHAT TYPES OF TESTS TO USE

Drug-testing technology is advanced and very accurate. The testing methodologies differ but can be summarized by reviewing a number of major scientific methodologies for detecting drugs in urine. Typical drug tests will show whether the person being tested has a drug in his or her body at the time of the test. The results of drug tests reflect the presence of drugs and drug metabolites. Virtually all intoxicating drugs of abuse can be identified with current technology.

Urine is the most appropriate body tissue for testing because urine is the body's natural waste-disposal product and consequently contains the residue of drug use. Drug substances are often present in higher concentrations in the urine than in blood, making identification of a drug substance easier in urine. The alternative body fluid used for drug testing is blood. In addition to the fact that drug substances are often present at lower levels in blood, blood testing is more difficult, more intrusive, more painful for sampling, making blood a less preferred body fluid for testing.

Drug tests will not document the time during which a drug was taken or necessarily the amount taken. Drug tests will answer one question: is a drug in the user's body at the time of the test?

Drug testing can be done relatively quickly and efficiently. Currently available tests using urine span a broad spectrum from preliminary screening methodologies utilizing immunological assays, enzymes, and radioisotopes, to gas chromatography/mass spectrometry (GC/MS) testing, which utilizes the most accurate quantitative detection system available. Tests generally can be identified in these categories: immunological assay, radioimmunoassay, thin-layer chromatography, gas chromatography, and GC/MS. There are other variations from these general testing methodologies, including liquid chromatography and high-pressure liquid chromatography. The drug-testing laboratories principally will utilize several of the methodologies just mentioned to provide drug-testing results to employers. What is most important is that a confirmation test of each urine specimen be performed.

The urine specimen in a drug test undergoes a preliminary screening. Either an immunological assay or radioimmunoassay is used for the preliminary screen by most drug-testing laboratories, although thin-layer chromatography identification of drug substances can also be used. If the urine specimen tests positive, a confirmatory test using an alternate methodology is appropriate and should be required. The most accurate and sensitive methodology is the GC/MS. Federal guidelines for drug-testing programs for federal agencies require that the GC/MS be the confirmation method, and a number of state laws are now specifying such confirmation tests as a requirement for the testing of employees in their states. The methodologies all seek to identify unique characteristics—metabolites identifiable within the urine that reflect specific drugs. Of concern to employees and employers alike are the potential for cross-reactivity or misidentification of such metabolites and the potential conclusion that someone may be using an illegal substance when in fact he or she is not but may be taking cough medications, antihistamines, cold tablets, or other substances. The GC/MS confirmation precludes the misidentification or cross-reactivity process that may become a problem in a preliminary immunological assay. The immunological assay and radioimmunoassay drug tests are 97 percent or more accurate, even with the possibility of cross-reactivity. When combined with the GC/MS confirmation, the testing process is 100 times more accurate than 99 percent—as the Mayo Clinic findings have indicated, it is virtually 100 percent accurate with respect to the identification of marijuana.

In addition to testing methodologies, however, drug-testing accuracy depends upon a clear chain of custody, notice, appropriate safeguards, record keeping, labeling, and the securing of frozen specimen after positive test results. The checklist that we have developed for standards and reliability of a drug-testing laboratory includes a number of requirements for the laboratory, as follows:

- —proficiency of personnel;
- —selection of equipment and procedures;
- —availability of an updated operating manual;
- —quality outside assurance;
- —accreditation and licensure by the state and by the Drug Enforcement Administration;
- —accurate storage of documentation and records;
- —frozen storage of positive specimens for one or two years;
- —availability for testimony in court or arbitration cases;
- —access to employers with respect to ongoing operations and testing results and procedures;
- —maintenance of confidentiality and security of drug-test results and testing records;
- —adequate backup of equipment and personnel to provide and sustain proficiency and accuracy at all times;
- —appropriate management direction by someone with a Ph.D. in toxicology and provision by such a qualified person of all drug-test results in writing; and
- —a proficiency testing program utilizing urine specimen sampling on a periodic basis to verify testing procedures.

COURT DECISIONS

The courts have tended to sustain drug-testing programs for employers that have done their homework. The principal federal appellate court decisions involving drug testing include determinations by the federal Courts of Appeals for the Fifth and Eighth circuits involving both public and private employers. The cases involve the U.S. Customs Service, Burlington Northern Railroad, and the Iowa prison system; all of them involve random testing of individuals. The appellate court overturned a lower court's decision in the Customs case and found that the Customs Service could provide drug tests for individuals about to be promoted into law enforcement positions of responsibility, even though there was not a nexus or suspicion of drug-use violation by these individuals. The Eighth Circuit found similarly that Iowa prison guards could be tested at random even though they neither had demonstrated aberrant behavior nor had been reported to have violated drug-use rules at the penitentiary. Courts found that the nature of their work required an extra measure of reliability and public confidence and that the expectation of privacy was outweighed by compelling public interest. The Eighth Circuit's appellate court also ruled in the case of Burlington Northern, the ruling being that railroad employees returning from furloughs could be subject to testing prior to their return to jobs on the railroad.

In general, the drug-testing programs that provide notice to employees, that spell out the parameters of discipline, and that are applied consistently with drug-testing procedures that take into consideration the necessary concern for accuracy, confirmation of test, clear chain of custody, and documentation of records have been supported. As important as the procedures and testing methodology are, the first and foremost need is to establish a policy under which the drug-testing program would operate. In most cases in which the courts have struck down drug-testing programs—the Plainfield Police case, for example—the company or agency has not provided notice, has not linked the drug test with the implementation of a policy, or has an unreasonable policy.

LEGISLATIVE ACTIONS

Approximately thirty states have been considering legislative standards or safeguards for drug testing. A number of states—nine at last count—have in fact passed drug-testing statutes. Generally, the statutes provide for confirmation of drug tests prior to action by employers, require prior notice to job applicants and employees alike before the imposition of testing, and establish other criteria related to the availability of testing results. The type of testing methodology that should be used for all confirmation testing is GC/MS. In Minnesota, a drug test may not be required for preemployment purposes unless an applicant has been conditionally offered a job position. In that state as well as in Connecticut, individuals tested have a right to receive a copy of their test results. Utah has implemented a pro-drug-testing statute that encourages employers, in effect, to test employees prior to employment, for fitness for duty and for cause, as well as on a periodic randomly announced basis under certain circumstances.

Legislative action will continue. To date it has been concerned with providing safeguards regarding procedures and access to information rather than with prohibiting drug testing per se. As additional case decisions emerge from both federal and state courts, legislative action will vary.

LABORATORY CERTIFICATION

The federal government does not have a designated federal drug-testing laboratory stamp of approval, although it intends in the near future to establish a certification program through the National Institute on Drug Abuse. What the federal government has recently issued, however, are guidelines for drug testing for federal agencies. A recent issue of the *Federal Register* specifies under what auspices and provisions federal agencies may test for drugs.[2] The criteria and specificity of these guidelines are both comprehensive and instructive. Federal agencies will be required to carry out confirmation testing on a mandatory basis, using the GC/MS method. The cutoff levels—those above which a drug will be found to be positive—identified record-keeping systems for drug-test results, and laboratory minimum standards are spelled out. The agencies must at least test for marijuana, cocaine, amphetamines, heroin, and phencyclidine (PCP). Specimen collection and chain-of-custody procedures are documented. A witness is not required to observe the provision of the urine specimen unless the agency has reason to believe it may be adulterated or tampered with. Instead of observed specimen collection, temperature, pH, and specific gravity are immediately determined and other safeguards such as a dry room and removal of bulky articles are stipulated. Guidelines undoubtedly will be subject to further revision and modification over time, but they represent an endorsement by the federal government of both

2. "Final Notice and Guidelines for Federal Drug Testing Programs," Department of Health and Human Services, in *Federal Register*, 52(157), 11 April 1988.

the need for drug testing and the need to do it correctly.

VALUE OF DRUG TESTING

The value of drug testing is severalfold: it is preventive, it is rehabilitative, it is reinforcing. Companies that have initiated drug testing along with education and training programs have experienced a decrease in accidents, a drop in absenteeism, and a significant curtailment in medical-benefit costs.

Employers will ask themselves, "Is drug testing effective and worth the effort and the potential employee resentment and misperception? Will it indicate impairment?" Most drug users and employers alike assume that once the high is gone, the drug is gone from the body and thus the impairing effects of the drug are also absent. The question arises of whether drug tests that show drug abuse at the workplace reflect past drug use off the job. The problem is that such a question misses the fact that drug substances stay in the body long after the high is gone. The concept of impairment from drug use is complex and basically undefinable in behavioral terms. Unlike alcohol, where telltale breath odor and lack of coordination and judgment are common manifestations of impairment, with drugs there is no readily observable marker, no predictable cause-and-effect time sequence impairment.

In a study by Dr. Jerome Yesavage, at Stanford University, 10 airplane pilots who had used marijuana after establishing a baseline of landing performance on a simulator indicated impairment from marijuana even 24 hours after use. These pilots each smoked a marijuana cigarette provided by the National Institute on Drug Abuse and then performed basic piloting skills on a computer cockpitlike simulator; then the deviations from the center of the runway, and comparisons with aileron and elevator movements were reported and recorded. Prior to inhalation of marijuana, the pilots deviated an average of 12 feet from the center of the runway. With the passage of 1 hour after marijuana use, the average deviation was 29 feet; 24 hours later, after a good night's sleep and no alcohol or other drugs, these same 10 pilots deviated from the center of the runway by an average of 24 feet, one pilot missing the runway altogether. The most significant finding in this study was that the pilots themselves thought that they were conducting their landing performances accurately, and research observers detected no signs or symptoms of physical impairment from their behavior. Thus drug testing may be the only method of preventing an accident from happening and of establishing a method to bring treatment to individuals whose drug use has not yet caused such a serious deterioration that physiological behavior cries out for intervention.

Most companies are now using standards of impairment per se. This means that a specific cutoff level of the drug in the urine is defined as signaling the potential of impairment in the same way that specific levels of alcohol in the breath or blood are defined as measures of impairment per se on the highway. The presence of an intoxicating drug in an employee's body fluid poses a risk and liability at work, a risk that need not be taken by the employee, employer, coworkers, or the general public.

A drug test performed on urine may show positive for several days after use of the drug, although this is principally true only of marijuana. The length of time a test may be positive will depend upon the drug used, its potency, the frequency of its use, and the amount of time since the last use. In general, the more recently used, the larger the amount used, and the more frequent the dosage, the longer the urine will test positive. Marijuana causes a positive test result for a long period of time while alcohol and cocaine, for example, make the test positive for short periods of time. These differences exist not because of the sensitivities of the tests but because of the differences in the way the body handles these drugs. THC, the active chemical in marijuana, is stored in the fatty tissues of the user's body, particularly the brain, for a long period of time. Measurable amounts of THC can be found for several weeks in the brain after a single use of marijuana. Alcohol, by contrast, is relatively quickly converted in the body to water and carbon dioxide, so alcohol tests will be negative within four or five hours of the last ingestion of alcohol.

One company that has initiated drug testing along with training in awareness has experienced a decrease in accidents, a drop in absenteeism, and a significant curtailment in medical-benefit costs. A company in the Chicago area, Commonwealth Edison, has had a drug-testing program for more than four years, and it provides a good example of how drug testing can work in conjunction with other initiatives. At Edison, accident rates and absenteeism have steadily decreased, and medical-benefit costs, which had been escalating at the rate of 26 percent per year, have leveled off. Edison's plan includes testing job applicants as well as employees for specific reasons, including reasonable suspicion, fitness for duty, and postaccident situations. The company has implemented an employee assistance program, providing treatment to individuals and their dependents with drug, alcohol, financial, legal, or other problems. The availability of this program serves as a counterbalance to the drug test and as a positive resource. An employee can use the employee assistance program confidentially, without risking suspension or, in some cases, termination of employment. At Edison, the drug test serves to protect the employer and coworkers by being a monitor on employees who have used drugs and have been through treatment and are returning to work. It may be used if employees are found to be in violation of company standards for fitness for duty before final termination of employment.

[Drug testing today represents an important tool in American industry's battle with the serious problem of drug abuse. Drugs today represent a drain on our industrial strength as well as a threat to the social fabric of our society. Drug abuse is a disease of denial that is neither victimless nor self-curing. While drug testing will not cure society's ills or completely eliminate drugs from the workplace, it is a significant deterrent that should be used with care and commitment, serving the interests of all.]

Employee Theft:
A $40 Billion Industry

By MARK LIPMAN and W. R. McGRAW

ABSTRACT: Employee theft is estimated to cost American business up to $40 billion annually. The costs to society include increased business failures, along with the resulting losses in jobs and tax revenues, and higher prices passed on to consumers. An intangible societal cost is the operation of the private justice system, completely separate from the criminal justice system. Because private employers focus on preventing crimes against their businesses, including employee theft, apprehending and prosecuting criminals ex post facto is rarely a priority, and prosecution is often avoided as troublesome and unsettling. Employee thieves generally do not feel any long-term commitment to the employer and are not motivated by financial need.

Mark Lipman is the founder of Mark Lipman Service, Inc., which is now owned by Guardsmark, Inc. He has over 50 years' experience as a private investigator, focusing on employee theft, and is the author of Stealing: How America's Employees Are Stealing Their Companies Blind. He presently serves as vice chairman of the board of Guardsmark, Inc.

W. R. McGraw started his career in 1955 with Mark Lipman Service, Inc., and he is presently a senior executive with Guardsmark, Inc. He is a member of the American Society for Industrial Security, a certified protection professional, and the author of several articles on security for national trade publications.

AMERICA'S national pastime is not baseball; it is theft. This is as true today as it was 15 years ago. If anything, the situation has grown worse. That might sound alarmist, but the facts cannot be ignored. In 1984, bank employees stole $382 million, nine times more than bank robbers stole.[1] The total lost by bankers to insider fraud and embezzlement rose to $850 million in 1985 and to $1.1 billion in 1986, and insider theft is a factor in about one-third of all bank failures.[2] Estimates of the cost of internal theft of all kinds range up to $40 billion a year,[3] and it is thought that from 5 to 30 percent of all business failures each year result from internal theft.[4]

SOCIAL IMPLICATIONS

The social implications of employee theft on this scale are enormous. At a time when major efforts are being made to understand and nurture the job-creation process, a large percentage of newly created jobs are being lost because of business failures that could have been prevented through improved security. If from 5 to 30 percent of all business failures each year are caused by employee theft, then it is reasonable to assume that the same percentage of new businesses—created through small business incubators, venture capital funds, and enterprise zones—will also fail because of employee theft. Inadequate security measures are causing millions of dollars, from both public and private sources, to be wasted.

Business and bank failures caused by employee theft also lead to loss of tax revenues, shaken faith in the stability of the financial system, and soured labor-management relations. While employee theft is accelerating, the growth of "intrapreneurship" in American corporations gives more employees access to corporate funds, goods, and information. The benefits of decentralizing decision making and diffusing responsibility throughout a corporation could be easily offset by the increased opportunities for employee theft. Ironically, the dismantling of organizational controls will require the imposition of heightened security controls, although these measures are obviously inadequate as they exist now.

Another business trend that could lead to an increase in employee theft is the downsizing of corporations. As corporations trim their payrolls, millions of employees see their friends and peers lose their jobs for financial or strategic reasons unrelated to job performance.

Feeding the surviving employees' sense of insecurity is the realization that flattening the hierarchical structure and streamlining the reporting lines lead to less mobility within the corporation. Promotion and advancement become more difficult. In a recent survey of middle managers, only 41 percent expected to be working for the same company in 5 years and only 17 percent expected to be there in 10 years.[5] A decline in company loyalty and long-term commitment can be expected to lead to an increase in employee theft.

---

1. U.S., Department of Justice, cited in *Security Systems Digest*, 17 Feb. 1986.
2. William Kronholm, "Inside Theft a Major Factor in One-Third of Bank Failures," Associated Press, 9 June 1987.
3. U.S., Department of Commerce, cited in Gloria Meinsma, "Thou Shalt Not Steal," *Security Management*, Dec. 1985, p. 35.
4. The low estimate of 5 percent is given by the National Council on Crime and Delinquency; the estimate of 30 percent comes from the U.S. Chamber of Commerce. Both are cited by Georgette Bennett, *Crimewarps: The Future of Crime in America* (Garden City, NY: Doubleday, Anchor Press, 1987), p. 105.
5. Stanley J. Modic, "Is Anyone Loyal Anymore?" *Industry Week*, 7 Sept. 1987, p. 83.

WHAT THEY STEAL

Employee theft comes in many forms, and it can be analyzed in various ways. Perhaps the simplest is the nature of the item stolen. Seen from this perspective, employees steal money, goods, services, and information.

Money

Stealing money can be as easy as dipping into the till or as complicated as embezzling funds through computer manipulation of accounts. Opportunities are rife, and the methods are limited only by the sophistication and imagination of the thief. An executive can pad an expense account and submit vouchers that have been falsified; a clerk can overcharge customers and keep the difference; accountants can make payments to imaginary businesses; and purchasing managers can accept bribes or arrange for kickbacks.

Computers have given thieves with the necessary expertise an entirely new avenue of access. According to a study by Ernst & Whinney, computer fraud alone costs American businesses between $3 billion and $5 billion each year.[6] Of all computer criminals, 99 percent are employees of the companies that suffer the losses.[7] In a recent survey, 70 percent of the respondents listed unauthorized use by employees as one of the three biggest threats to their companies' computer operations. Only 24 percent named use or misuse by outsiders as a major threat. One in six of the respondents said their company had detected a computer crime sometime in the last five years. Larger companies with more employees who work with computers are more likely to find their computers used against them. Of companies with more than 1,000 employees, 32 percent have detected incidents of computer crime. The rate rises to 47 percent for companies that have at least half of their employees working on computers.[8]

Goods

Internal thieves steal anything that is not tied down, and almost anything that is. Tools, equipment, raw materials, and finished products are the major categories of the stolen goods. Wrenches, earth-moving equipment, lumber, toasters, and anything else that people either want themselves or can fence are being stolen by employees. According to Arthur Young & Company, which conducts an annual survey on shrinkage losses for the International Mass Retail Association, retail organizations—mass merchandisers, department stores, and specialty shops—attribute more of their inventory losses to employee theft than to any other factor. The best estimate of retailers is that, in 1985, employees were responsible for 43 percent of the losses, shoplifters for 30 percent, bookkeeping errors for 23 percent, and vendor theft for 4 percent. The total shrinkage for the 113 companies that responded to the survey came to $1.283 billion, or 1.8 percent of sales.[9] Extrapolating from these figures, some analysts claim that retailers nationwide lose $20 billion annually to employee theft.

Services

Employees also steal services or time. What used to be called goofing off is becoming so prevalent that it is now a

---

6. Peter H. Lewis, "When the Password Is a Passkey," *New York Times,* 27 Sept. 1987.
7. Ibid.
8. "How Business Battles Computer Crime," *Security,* Oct. 1986, pp. 54-60.
9. "Eighth Annual Security and Shrinkage Study" (Survey, Arthur Young & Company, 1987).

focus of attention. Robert Half, of the executive-recruiting firm Robert Half International, conducts annual surveys of time theft. He has estimated that employers lost $150 billion in 1984 to time theft.[10] Coming to work late, taking excessive breaks, conducting personal business, taking long lunch breaks, and leaving early all add up over a year's time.

Falsifying time cards is another method of stealing time. A major study of employee theft found that 5.8 percent of retail employees admitted receiving more pay than warranted by the number of hours worked. For hospitals, the rate was 6.1 percent, and for manufacturers the rate was 9.2 percent.[11] Falsifying time cards can also involve management, as in the case of a manager's borrowing $25 from a subordinate and crediting the latter with $30 worth of overtime.

Personal use of long-distance telephone lines also mounts up, as the U.S. government has discovered. Unauthorized use of the federal long-distance telephone system inflates the phone bill by $89.5 million a year. The work time that employees spend making those personal calls costs an additional $76.3 million.[12]

Information

Competitive information is a commodity like any other, and it is regularly bought and sold. Customer lists, marketing and sales data, strategic plans, unique manufacturing processes, and research and development breakthroughs are all examples of trade secrets that unscrupulous competitors are willing to purchase. Gaining access to such secrets usually requires the collusion of an insider.

It is often no longer necessary for corporate thieves to pack their briefcases with printouts, memos, and reports. A floppy disk will suffice. With increasing numbers of executives using home computers and modems to communicate with their office computers for evening and weekend work, taking home floppy disks is not seen as suspect but as a sign of dedication and ambition.

Competitive bid information is like a gold mine to the corporate thief. Often a company does not even suspect that its secrets are being sold until a consistent pattern of loss emerges. Such was the case with an electronics company that time after time found itself underbid by a particular competitor. The company lost $20 million in contracts until an undercover investigation discovered the source of the problem.[13]

Insider theft of company trade secrets is a dismayingly gray area when the purported thief has not sold the secrets but has changed his place of employment, going to work for a competitor or even for himself. To what extent may the ex-employee use for the benefit of the new employer or for his own benefit the information he was privy to in his previous employment? Such information may be of limited and temporary value—for example, advertising plans—or be of a permanent nature, such as unique technological or manufacturing processes. If the latter is the case, then the competitive use of the information could threaten the very existence of the victimized

10. "High Cost of Time Theft," *Dun's Business Month*, Jan. 1985, p. 23.
11. Richard C. Hollinger and John P. Clark, *Theft by Employees* (Lexington, MA: D. C. Health, 1983), p. 42.
12. Sydney Shaw, "New Rule Would Crack Down on Federal Phone Abuse," United Press International, 27 Mar. 1987.
13. Richard Morais, "Sam Spade Goes Corporate," *Forbes*, 25 Feb. 1985, p. 126.

company, if it is not sufficiently diversified.

In recent years, there have been numerous cases of this sort. One of the best known involves IBM and five ex-employees who left to set up their own company. The new company was to compete with IBM in making thin-film read-write heads for computers. IBM claims that it spent $200 million over almost one and a half decades to develop the process that Cybernex, the new competitor, developed in months. IBM filed suit against the new competitor for patent infringement; Cybernex countersued for restraint of trade. In the settlement, IBM reportedly received everything it had asked for, including reimbursement of legal costs and salaries it had paid the founders of Cybernex during the time they had spent planning their venture while still employed by IBM.[14]

Cases involving purported theft of trade secrets are often not cut-and-dried. The line that separates theft from commendable entrepreneurship is blurred. The increasing inclination of employees to go their own way, however, regardless of how commendable it may be, is related to the general decline in company loyalty.

How many employees steal from their employers? Experienced private investigators believe that one-half of the work force engages in petty theft from employers; taken are paper, pens, and other small items. Of those 50 percent, however, half also steal important items, and from 5 to 8 percent of all employees steal in volume.

A study by the U.S. Chamber of Commerce found that up to 75 percent of all employees steal at least once and up to 40 percent at least twice.[15] Another study, which guaranteed anonymity to respondents, reported that approximately one-third of employees admit to stealing in various ways, ranging from taking merchandise to abusing reimbursement accounts. For manufacturers, the self-reported figure was 28.4 percent; for hospitals, 33.3 percent; and for retail organizations, 35.1 percent.[16]

GREED, NOT NEED

Why do so many employees steal? Much attention has recently focused on the need for ethics, both corporate and personal, in business. Companies that intentionally dump toxic wastes, knowingly market defective products, or engage in corrupt bidding practices may be exceptions, but their conduct is glaring enough to raise real concerns about how ethical considerations enter into business decisions. On the personal level, the insider-trading scandals on Wall Street have given added impetus to consultants and universities to offer seminars and courses on ethics for business executives.

The concern for ethics in the business life of our nation must not be restricted to the relatively few whose schemes produce headlines. While we worry that the apex of the pyramid is showing signs of wear, the base has been corroding for years. When 43 percent of Americans between the ages of 15 and 19 admit that they have shoplifted, the problem touches more than retailers.[17] Cheating has be-

14. David E. Sanger, "I.B.M. Wins Patent Settlement," *New York Times*, 8 July 1986.

15. August Bequai, "Trusted Thieves," *Security Management*, Sept. 1986, p. 83.

16. Hollinger and Clark, *Theft by Employees*, p. 42.

17. George P. Mochis, "A Study of Juvenile Shoplifting" (Paper, Department of Marketing, Georgia State University, n.d.), cited in Lewis H. Lapham, Michael Pollan, and Eric Etheridge, *The*

come so endemic in American life that allegations are met with a shrug and a grin, as both pitchers and hitters exemplified during the 1987 baseball season.

Far too many companies treat employee theft as simply another cost of doing business. The costs are passed on to the consumer; the thieves—when they are caught—are dismissed instead of prosecuted; and fear of costly judgments from defamation lawsuits often prevents companies from giving the thieves the bad references they deserve. Business does not bear the blame alone, for our society as a whole does not wish to pay the costs of prosecuting and punishing all those who are caught by private and proprietary security.

Employees who steal from their employers do so for numerous reasons, but economic necessity is rarely one of them. An employee who becomes involved with drugs or gambling may resort to stealing to pay for his or her habits, but lack of money is not the problem. Economic pressure explains only a small proportion of employee thefts. Recent research does not support the theory that employees steal out of necessity.

The researchers did, however, find a correlation between lack of company loyalty and theft tendency. Employees who were unhappy in their jobs or who felt dissatisfied or resentful because they thought they were not treated fairly were the most likely to steal from employers. Feelings of personal fulfillment and general job satisfaction weighed even more heavily than the wage level. Employees who felt a long-term commitment to the employer were less likely to steal than those without such a commitment. Younger employees were also more likely to steal, but this tendency is attributed to the fact that young employees are less committed to the company.[18]

The study's findings on theft by young employees are supported by the experience of Service Merchandise Company in 1984. Of the employees caught stealing that year, 52 percent were under the age of 21 and only 3 percent were over 40 years of age. One-half of all those caught stealing were caught within four months of their hiring. An interesting aspect of this study is the finding that more than 90 percent of those caught evaluated their relationship with their supervisors as "above average." The vast majority— 95 percent—of the supervisors' evaluations of these employees were "average," "good," or "excellent."

Why did they steal? The major reasons given were "need" or "greed" (52 percent); "no reason" or "didn't know why" (16 percent); and peer pressure, opportunity, personal problems, or because they thought it was easy (16 percent).[19]

THE PRIVATE JUSTICE SYSTEM

Employee theft is a major causal factor in what has come to be called the private justice system. The system arises from the various ways in which private businesses handle crime without resorting to the criminal justice system. For a number of reasons, a business firm might prefer to deal with criminals— internal or external—in its own way. The decision of whether or not to prosecute takes many factors into account. A longtime, faithful employee who embezzles to cover a spouse's medical costs

---

Harper's Index Book (New York: Henry Holt, 1987), p. 81.

18. Hollinger and Clark, Theft by Employees, pp. 63-68.
19. Michael H. Miller, "Portrait of a Thief as a Young Man," Security Management, June 1986, pp. 38-42.

from a catastrophic illness or a teenager with an otherwise excellent work record might well be dismissed or reassigned rather than handed over to the police.

Central to the private handling of employee theft is the focus on loss prevention. Private security is employed to prevent crime, not to act as an extension of public law enforcement agencies. Many firms are well satisfied if they can discover the thief's method and alter the security operations or procedures to ensure that the same method will not succeed in the future. Pursuing prosecution of the offender could envelop the firm in drawn-out and time-consuming complications: legal staff would have to be consulted, evidence gathered, depositions taken, employees might have to take time from work to testify, and so on. The most efficient result for the firm in many cases is a confession, an agreement for restitution, and enhanced security procedures.

A recent survey indicates the relative weights accorded to reasons for not involving the criminal justice system in cases of employee theft. Of the respondents, 70 percent said they did not report some crimes because the loss was considered minimal. There were, however, 54 percent who sought restitution in cases that were handled internally; apparently those losses at least were not too minimal. Incomplete evidence was given as a reason by 59 percent, and avoidance of publicity was the reason for 31 percent. Avoiding trouble and expense was cited by 26 percent, and another 23 percent noted problems with the justice system.[20]

There are a number of drawbacks to the private justice system and the way it operates. In functional terms, it carries little deterrent value because punishment is so light and there is little certainty of being caught. The entire thrust of the system is not to catch and punish the criminal but to prevent the crime.

An individual business firm or an industry as a whole can justify the operation of the private justice system, but for society, there are moral and political dimensions that call the system into question. Suspected and confessed thieves are treated differently by the two systems. Rules of evidence and procedure, methods of investigation, and punishments differ greatly.

The existence of the two systems raises fundamental questions of fairness. For example, the *Miranda* rulings apply only to government agencies and not to private employers. In a private investigation, suspects do not have to be warned that what they say may be used against them. Evidence that clearly indicates criminality but that might not stand up in court may be used in a private investigation to coax a confession from a suspect.

The existence of two systems of justice leads in effect to selective prosecution. The price of embezzlement for the employee of a firm using one justice system might be years in prison, while the employee of a firm operating under the other justice system, although equally guilty, suffers only the loss of employment.

Such a result is certainly not socially desirable, but society has apparently decided that paying for the prosecution and incarceration of even violent offenders is becoming less desirable. Court dockets are already full, and overcrowded prisons are under court order to reduce their population. What would the overburdened criminal justice system do if thousands of employee thieves were prosecuted?

---

20. "Inside the Private Justice System," *Security*, Aug. 1987, pp. 44-49.

CONTROLLING EMPLOYEE THEFT

Employee theft cannot be eliminated, but it can be controlled by good management techniques and judicious, intelligent application of acceptable security procedures.

The first control point is the most obvious: the personnel selection process. Fortunately, honest people for the most part stay honest; unfortunately, dishonest people tend to stay dishonest. The past turns out to be the best predictor of the future. Employers who fail to conduct background investigations, to check—as far as possible—previous employment records, and to carry out in-depth interviews should not be surprised at their high levels of employee theft.

Once hired, employees must be made to feel that they are part of the firm and that their contributions are important in determining the firm's success. Employees who are disgruntled and disaffected represent the greatest threat of theft and vandalism. Employees who feel abused or exploited tend to get even in whatever way they can. Havoc can be wreaked by disgruntled employees, such as the accountant who changed payroll records and gave everyone in a particular company a raise. Modern payroll checks are so complicated and impersonal that the changes were not obvious, and if there were any employees who did notice, they did not bother to thank management for the unannounced raises. The company was losing heavily enough to begin an investigation of employee theft. After no evidence of the theft was found, it was the disgruntled-employee theory that led to the discovery of the payroll scheme.

Good management practice with respect to employees' feelings also requires paying well. The most important control against employee theft is a good salary or wage. It will not stop the confirmed thief, but there is no doubt that more stealing goes on where employees are underpaid than where they are overpaid.

The third component in a complete plan to reduce employee theft is an unequivocal company policy or code of ethics that is impressed upon all employees at an initial orientation. The policy must clearly indicate what ethical standards are expected of the employees, as well as the consequences of violating those standards. The penalties for violating the policy must not be applied selectively, and the company must itself operate according to a high ethical standard.

Most employees are honest, and it is the responsibility of management not to offer them temptations that are beyond their power to resist. Our consumer society assaults people every day with powerful messages created to stimulate their acquisitiveness. To tempt employees constantly with easy opportunities for theft is irresponsible.

Finally, reducing theft requires the implementation of adequate security measures. The security measures are available, from sophisticated computer security systems to undercover investigations. Awareness of the problem of employee theft is spreading, and more companies are recognizing that handsome returns can be earned from modest investments in security. In 1974, only 10 percent of the *Fortune* 1000 companies used undercover security agents. By 1985, the number had increased to 50 percent.[21]

Employee theft is a $40-billion-per-year disease eating away at the health of

21. *Security Letter*, [1986], cited in Lapham, Pollan, and Etheridge, *Harper's Index Book*, p. 83.

American companies. The symptoms are masked because so few incidents ever see the light of day and because the social costs are widely distributed. Those costs are real, however, and they include business failures, lost jobs, lower tax revenues, and higher prices passed on to consumers.

Civil Aviation: Target for Terrorism

By WILLIAM A. CRENSHAW

ABSTRACT: On a global basis, few major industries have been as affected by the growing menace of terrorism as has civil aviation. This article explores the expanding role of private security in protecting the domestic air transportation system against the threat of terrorism and how that role is being performed. Concern is increasing that terrorist attacks of the type that have occurred in the Middle East and Europe may gravitate to the United States. In that event, the response would include immediate and significant requirements for additional private security personnel and equipment, possibly on a level comparable to the implementation of mandatory screening in 1973.

William A. Crenshaw received his Ph.D. in international affairs from the University of Miami. While a career military intelligence officer, he served as military attaché to a Latin American country and as special assistant to the director, Defense Intelligence Agency, for the National Foreign Intelligence Board. A certified protection professional, he is a consultant and security adviser to multinational clients.

THE rate of the growth of private security in the United States over the past several decades has been phenomenal. More than 1.5 million persons are engaged in private-security-related activities, nearly twice the combined personnel of federal and local law enforcement agencies. Technical security services offered by private sources have become increasingly available in an expanding market. Security sales are brisk, and the pace of technological progress in the development and improvement of security equipment and systems is impressive. These trends appear likely to continue.

In large part, the rapid expansion of private security services is in response to the sharp increase in the incidence of common crime that has occurred over time in this country. As common crime has grown faster than have public law enforcement resources to combat it, the American public has become progressively more crime conscious. A second and even more dramatic impetus, however, terrorism, likewise has contributed to the explosive expansion of private security.

TERRORISM: THE THREAT

Terrorism has seized and, temporarily at least, dominated world attention repeatedly in recent years. It has come to be widely considered a serious world problem.

An extremely emotional issue, terrorism has yet to be defined in a universally accepted form. More than 100 definitions have been developed by respected academicians, political organizations, and government agencies. A major area of controversy is the distinction, if any, that exists between politically motivated violence and that generated by common criminal activity. Opinions vary widely and are strongly held.

Dr. Brian Jenkins expresses a broader view, which I share. He defines terrorism by "the quality of the acts, not by the identity of the perpetrators or the nature of their causes. All terrorist acts are crimes."[1] Although more commonly used within a political context, terrorism is a tactic of intimidation that can be and has been employed by political extremists, common criminals, or unstable persons.

Academic interest in terrorism has increased markedly in recent years, but there yet appears an inadequate appreciation for the importance of the victim's perspective in assessing particular acts or threats of violence as terroristic or not. Certain individuals, groups, and nations are affected more than others by the threat of terrorism. Under different circumstances, even these same potential victims may be significantly more or less sensitive or vulnerable to terrorist threats. It is suggested that a victim's perception of the threat of terrorism also may be adversely affected by repetitive incidents of violence that individually may not necessarily demonstrate a sophisticated and deliberate intent by the perpetrators to influence a larger indirect victim group.

Simply put, the extent to which terrorism is effective depends in large degree upon the severity of the threat as perceived by the potential victim or victims. This perception may or may not reflect an accurate assessment of the terrorists' capability. Of consequence for private security, the response of the potential victim to that perception provides a

---

1. Brian M. Jenkins, "Statements about Terrorism," *The Annals* of the American Academy of Political and Social Science, 463:12 (Sept. 1982).

measure of the effectiveness of the terrorist threat.

As violent as a terrorist incident may be, it is, in the final analysis, the threat of recurring violence, not past actions, that prompts the continuation or intensification of protective measures. Prevention, or at least moderation of the impact of potential incidents, is a role well suited for private security in complementing normal law enforcement and other government agencies. The extensive requirements for developing an effective defense against terrorism, in fact, necessitate the participation of private security.

The role of private security in countering terrorism is already well established in the United States and is apparent to the general public in the increased protection for certain government buildings and public areas. Less visible but more intensive is private security's involvement in the protection of nuclear power plants. But even more dramatic is the large and highly visible role of private security in the protection of civil aviation against the threat posed by terrorism.

CIVIL AVIATION: TARGET FOR TERRORISM

Modern terrorism has seriously affected international civil aviation and has affected it more, on a global scale, than any other major industry group. Despite statistical evidence that attacks against commercial air travel represent only a small percentage of total terrorist acts reported, there appears an unusual quality about civil aviation that, at least at the time, galvanizes public attention to focus on an aerial hijacking or related incident more than violence in a more static environment does.

The impact of terrorism upon the commercial air transport system is revealed clearly by the immensity and scope of the response by the industry, governments, and the general public. The social and economic costs of protection against the perceived threat have been high.

AN EVOLVING THREAT

An ominous trend began with the first hijacking of a U.S. commercial airliner near Key West, Florida, in May 1961. Gradually increasing in number over the next few years, hijackings spurted to epidemic proportions in 1968 with 23 attempts made against domestic flights. In 1969, aircraft hijackings in the United States reached a record 40 attempts.

Initial efforts to protect commercial aviation and the traveling public relied heavily upon a process of selective screening of suspect passengers that did not prove effective in preventing further hijackings. At that time, very few domestic incidents individually fit neatly into a category of conventional terrorism, but the cumulative incidents created widespread public alarm and provided the impetus for a massive security program that generated major requirements for security personnel and equipment.

In January 1973, the United States implemented a controversial, nationwide program requiring the security screening of all commercial aircraft passengers and their carry-on articles. There was less than unanimous support for the ambitious policy; critics had raised serious reservations concerning the physical ability to conduct such an ambitious screening program competently and the large costs required. Personnel and electronic screening requirements to operate such a system were extensive. Essential to the success of the program was the ability to conduct the security screening rapidly to avoid delays in check-in and boarding. The concept of a sterile, or screened, preboarding area resulted. Reinforced by the short-lived bilateral antihijacking agreement reached with Cuba in February of that year, the program had almost immediate results, with hijacking attempts falling from 30 in 1972 to 3 in 1973.[2]

The implementation of a mandatory search of all passengers and their carried possessions required a fundamental shift in the conservative concepts of privacy to which the American people, perhaps more than many national groups, were accustomed. Considering the intensity of these long-held values, the ready acceptance of the restrictions and impositions of the security inspections was even more indicative of the extent of the concern of Americans over the growing threat of violence against commercial air travel.

In a second outbreak of hijackings, which peaked in 1980, the perpetrators exploited weaknesses in the screening program to smuggle substances on board aircraft that were, or were claimed to be, inflammable or explosive. More recently, public concern for the security of domestic air travel has become overshadowed by a growing preoccupation with the international threat of terrorist attack, primarily in the Middle East and Europe, against American and foreign citizens, airliners, and airports. This concern was reflected in a major drop in air travel to those areas in 1985, with resultant significant economic impact on tourist-related industries. This international threat remains, but a question of urgent and critical importance for the United States, and for the impact on private security, is whether this form of terrorism will eventually expand to the United States.

THE TASK

Protecting civil aviation against hostile threat is an immense task. American commercial airliners are involved in more than 5.5 million flights annually. Nearly 2 million passengers are transported daily. The provision of protection for aircraft, airports, and other ground facilities requires an extensive outlay of labor and capital as well as extensive interaction between government agencies, the air transport industry, and pertinent service industries.

Under the regulation of the Federal Aviation Administration (FAA), air carriers are responsible for conducting security screening of all passengers and cabin baggage carried on their respective flights. This includes payment for the cost of law enforcement support of the screening process. Airlines are also charged with controlling access to and movement within their "exclusive areas" at each airport. Airport operators are responsible for providing protection for the terminal and other nonexclusive operational areas.

PASSENGER SCREENING

The passenger and carry-on baggage screening process is the most visible element of the overall security program and, as such, greatly affects the public's perception of the effectiveness of security measures. The program is expensive. Approximately $200 million, or around 40 percent of total annual security-

---

2. U.S., Department of Transportation, Federal Aviation Administration, "U.S. and Foreign Registered Aircraft Hijackings Statistics 1961 to Present," updated 1 Jan. 1986, pp. 1-134.

related costs for domestic air transportation, are allocated to this program.[3]

Since the total screening policy was implemented in 1973, security personnel have checked well over 7 billion persons and more than 9 billion carry-on articles.[4] Increasing air travel is rapidly generating additional screening requirements. Approximately twice as many persons were cleared through security checkpoints in airports in 1986 as in 1981, 1.13 billion compared to 598 million, respectively.[5]

More than 1200 screening stations are operating in nearly 400 airports under U.S. control to carry out the program. The work force provided by private security is considerable; more than 6000 private security employees are directly engaged in screening activities alone. Some 1500 law enforcement officers are directly engaged in support of screening activities.[6]

Although the configuration may vary depending upon aircraft layout and screening volume, a typical high-volume screening station is equipped with one or more metal detectors and X-ray machines operated by a trained security team. Refinements have been made over the 15 years that the mandatory screening program has been in effect, but the basic technology of the security screening station has not changed markedly.

3. Interview with Eastern Airlines security executive, Miami, FL, 27 June 1985.
4. U.S., Department of Transportation, Federal Aviation Administration, *Semiannual Report to Congress on the Effectiveness of the Civil Aviation Program: January 1-June 30, 1986* (Washington, DC: Department of Transportation, 1986), pp. 3-4.
5. Ibid., p. 15.
6. Air Transport Association, *Protecting Air Transportation from Hijacking and Sabotage* (Washington, DC: Air Transport Association, 1985), p. 1.

More than 1000 X-ray machines are being utilized for the security screening of articles carried on board by passengers. This equipment has proven highly effective in weapon detection. Based on past experience, approximately 95 percent of the smuggled firearms or other weapons seized at screening stations are detected initially with this equipment. Current-generation equipment in use at a station depends totally on operator proficiency for the identification of suspect shapes and densities for a more intensive manual search.

More than 2000 walk-through magnetometers, or metal detectors, are installed at airport screening stations, with 2 to 3 percent of the weapons seized at security screening stations initially detected with this equipment. The large spread in detection rates between X-ray and metal detection equipment may depend more upon violator behavior than the screening equipment.

Automated security screening of passengers and carried articles is essential in processing the high volume carried by commercial aviation, but it cannot eliminate the requirement for manual search on an exception basis. Security personnel conducting physical searches have uncovered approximately 2 percent of the concealed weapons seized at screening stations to date. In view of weaknesses in automated detection of explosives and inflammables, manual search remains of extreme value and has been reemphasized in recent years.

The probability that an actual weapon will be detected during the screening process has proven very small. Of the over 7 billion passengers and over 9 billion articles of cabin baggage screened since the program was implemented, around 40,000 firearms or other weapons have been found and some 13,000 persons arrested as a result. Obviously, not all persons found in possession of a weapon had intended to use them to seize control of the plane.

Effective training and testing are essential to ensure the competent operation of such a labor-intensive program. Security screeners complete a basic course of instruction in equipment operation and in the performance of screening procedures. This brief preparation is supplemented by on-the-job training and formal refresher training. As travelers have frequently noted, the performance of individual security screeners may vary widely due to ability and motivation.

Screening performance is evaluated by unannounced tests conducted by the contractor, airline, and FAA inspectors. In a series of random tests in 1986, FAA inspectors attempted to smuggle over 2400 mock guns and other weapons aboard aircraft at various airports. An average of only 80 percent were detected by the security screening process. The performance at individual airports varied widely, however, with Anchorage, Alaska, and Miami catching a high 99 and 98 percent of the attempts, respectively, compared to 34 percent at Phoenix.[7]

HAS SCREENING WORKED?

Spurred by such reports, questions have been raised from time to time concerning the effectiveness of the domestic screening program and the justification of its expense. Admittedly, the cost of prevention is high. The FAA claims that at least 117 hijacking attempts have been prevented by mandatory screening procedures in the United States.[8] Considering the screening costs over the past 15 years, a thwarted hijack costs millions of dollars.

The domestic threat has been sharply focused. Only 46 of the nearly 400 airports under U.S. control have been points of departure in 97 hijackings attempted since the implementation of total passenger screening in 1973.[9] Hijacking incidence is even more sharply concentrated among that small group. Hijackers boarded at Miami on 15 occasions and a combined 13 times at New York's Kennedy and La Guardia airports. These 3 airports accounted for 29 percent of total reported incidents. San Juan, Puerto Rico, was the point of boarding for six attempts, followed by Los Angeles and Chicago for five and four incidents, respectively. Tampa, Atlanta, and Denver experienced three incidents each. Hijackers boarded at 6 airports on two occasions each and once from 32 others.

A valid assessment of present need cannot be based solely on past experience, however. Whether totally effective or not, passenger security screening has developed a degree of permanency that would be difficult to roll back selectively at low-incidence airports without creating a negative public perception or encouraging a shift, over time, in security-related episodes to those locales. It is the future threat, rather than past incidents, that necessarily must determine the proper course. A point of major concern is that, within the United States, the security protection of civil aviation, as

---

7. "Miami Airport Security Scores High," *Tampa Tribune,* 23 June 1987.

8. U.S., Department of Transportation, Federal Aviation Administration, *Semiannual Report to Congress on the Effectiveness of the Civil Aviation Program: January 1-June 30, 1985* (Washington, DC: Department of Transportation, 1985), p. 3.

9. U.S., Department of Transportation, Federal Aviation Administration, "U.S. Registered Aircraft Hijacking Statistics 1961 to Present," updated 1 Jan. 1987, pp. 3-6.

extensive as it is, has not been challenged to date by determined terrorist groups of the sort that breached the security of airports in Athens, Rome, and Karachi. It cannot be assumed that the results would have been substantially different had similar attacks been launched against Kennedy, Miami International, or other major U.S. airports.

The airlines, individually and through the Air Transport Association, are demonstrating a vital interest in strengthening security measures. Included on the Air Transport Association's safety agenda for 1987 was the stated intent to improve the quality of the airline passenger screening process and enhance public confidence in the adequacy of airline security. A probable outcome is greater direct participation by airline operators, possibly jointly through the Air Transport Association as well as individually, in the screening process and other areas that have previously been left largely to their security contractors. Pan American Airlines was the first to announce a proprietary security organization in efforts to increase control over international and domestic operations. American Airlines also has enlarged its company security force.

Already, airlines are moving to impose additional restrictions on the amount of hand baggage allowed in the cabin. Based primarily upon other safety considerations, such a policy has positive security implications in reducing both the screening load and the options for smuggling weapons for hijacking.

SYSTEM WEAKNESSES AND THREATS

The rapid progress of high-stress plastic and ceramic weapon technology presents a serious challenge to the continuing effectiveness of automated screening for weapons. Contrary to initial claims and concerns, the Glock-17 plastic pistol now available contains metallic components of a cumulative weight that can be detected by current-generation metal detectors if they are properly tuned, but it is indicative of the rate of technological advance. The true nonmetallic weapon will not be detected by the passenger screening equipment now in operation. X-ray imaging equipment, however, to the extent that it can discern shapes and densities, will be able to detect any completely nonmetallic weapons contained in carried items.

The most critical and immediate requirement for effective screening, however, is the detection of explosives and inflammables. A reliable automated capability for detection of these substances has yet to be achieved on the scale required by the massive volume of passengers and carried baggage. The second wave of domestic hijackings, reaching 20, or one-half of total worldwide incidents in 1980, was facilitated by the screening program's inability to detect inflammable substances that could be used in reinforcing a hijacker's threat. The increased incidence of terrorist bombings provides an even stronger incentive for developmental efforts in explosives detection. Several concepts, in varying stages of development, appear to have particular potential.

RESEARCH AND DEVELOPMENT

One concept under development is that of vapor characterization. It relies upon the detection of minute amounts of vapor by sampling, or sniffing, the surrounding air and can be used for screening persons or articles. Some electronic explosives detectors based on this principle are currently in production and have been relatively effective under closely controlled conditions, but they have yet to achieve the level of confidence required for sustained operation under the demanding conditions—high volume and rapid processing in an open environment—encountered in most passenger screening situations. Machines are undergoing field testing and should be ready for limited operational use later in 1988.

Effective explosives detection will necessarily slow the passenger screening process. It is anticipated that a sensing cycle of approximately six seconds per person will be required when this capability is developed and deployed.[10]

Another enhanced explosives detection capability is based upon tightly focused X-ray beams and automatic computer-based analysis of the size, shape, and density of the contents of parcels or luggage being inspected. This represents a significant technological improvement over existing X-ray inspection equipment, which is totally dependent upon operator proficiency for detecting suspect items.

A third concept concentrates upon detection of anomalies, possibly employing infrared, ultrasound, or dielectric analysis. An envisioned system would emit an alarm and prompt a manual search if an unprogrammed mass and density were encountered in automated scanning. A prototype explosives detector based upon thermal neutron activation also is under development for possible utilization in baggage or cargo screening.

CHANGING FORMS OF THE TERRORIST THREAT

Terrorist attacks against civil aviation have taken the form of hijacking, place-

10. Interview with Dr. L. O. Malotky, Federal Aviation Administration, Washington, DC, 12 Jan. 1987.

ment of bombs on aircraft or in ground facilities, firing upon aircraft in the air or on the ground, and armed attack on airport terminals. As passenger screening has improved at international airports in areas of higher terrorist activity, the nature of the threat has changed in part. Hijackings pose a continuing threat, but the use of explosives to destroy an aircraft in flight—rather than the use of a weapon for hijacking—has become a serious threat to international civil aviation.

The extent to which this form of threat may develop domestically affects both passenger screening and other aspects of air travel. The urgent necessity for an operational high-speed explosives detection capability encompasses baggage and cargo screening as well as the screening of passengers and carried items. An increased requirement for security personnel, services, and equipment already is apparent in augmented measures intended to avert such an attack.

The placement of bombs on board aircraft is not a new development. At least 90 deliberate explosions have occurred on civil air carriers, resulting in the deaths of over 1500 persons. U.S. aircraft have been involved in 15 such incidents since 1955. What is significant is the apparent intensification of that threat, augmented by the increasing sophistication and availability of newer explosives and detonation techniques.

Checked luggage and cargo

Discontinuance of curbside baggage check-in for international flights is one step noted by travelers as a measure against possible explosions in flight. A program is under way, both internationally and domestically, to ensure a positive match of passenger and baggage. If a checked passenger does not board the aircraft for any reason, his or her baggage is pulled from the flight. Increasing computerization should facilitate this matching and reduce unnecessary delay in departures of aircraft resulting from search by ground crews for a missing passenger's luggage. Special screening procedures are in effect for flights considered high risks. These procedures may include X-ray inspection of all checked baggage. A checked-baggage profile also may be employed in subjecting suspect baggage to X-ray or other types of inspection.

What about nonpassengers?

Passengers are not the only source of potential threat. Others, relying on authorized or unauthorized access to preboarding or airside areas, have varying opportunities to avoid screening and access controls in order to assist a passenger hijacker, to place an explosive, or to take other actions. The attempted hijacking of a Pan American jet at Karachi, Pakistan, in September 1986, for example, was launched by terrorists who gained access to the airport operations area by posing as security guards.

Considering the extensive activity in a modern airport operations area by employees of multiple airlines, the airport operating authority, and various permanent and temporary contractors, the control of access and internal movement is a challenge at best. Difficulties experienced at certain airports in controlling drug smuggling and cargo theft suggest the potential for terrorist exploitation as well.

The control of access to and movement within restricted or nonpublic portions of the airport is accomplished through the administration of an airport pass or pouch program. A see-and-challenge procedure is used by security, supervisory, and other personnel, controlling internal movement with varying effectiveness.

Common to card identification systems, the potential for misuse of stolen, lost, or counterfeited airport badges is a continuing concern. Anticounterfeiting protection increasingly is being enhanced by state-of-the-art measures. A recognized vulnerability, the pass exempts the wearer from passing through the screening process on the terminal side of most airports.

Security concerns about possible terrorist connections, raised in part by media exposures of hiring practices, have generated an increase in employee background investigations within the industry. A background or historical check is required of employees of FAA-regulated companies or agencies hired after November 1985. The depth of the investigation is limited, however, and, in the case of resident aliens, only covers from the green card forward. Terrorist connections would not be difficult to conceal.

Airport perimeter and internal protection

The trend is continuing for greater reliance upon electronic security equipment and systems for perimeter and internal control in airport operational areas to supplement patrols of police or private security officers. This trend can be expected to intensify with technological advancement and if and when an increased threat becomes apparent. In recent years, the air carriers have sharply increased private security requirements for protection of parked aircraft and pertinent ground facilities.

International security

Screening procedures developed in the United States in response to a prolonged rash of hijackings in the late 1960s and early 1970s became the basis for a worldwide standard through the sustained efforts of the International Civil Aviation Organization, national governments, and others. As a result of subsequent terrorist threats, however, especially in Europe and the Middle East, security measures currently in effect at many international airports are far more extensive than those in the United States. So, too, are specific measures being taken by U.S. and other airlines operating in those areas. These afford a valuable opportunity for developing and evaluating domestic options should the terrorist threat intensify within the United States.

More apparent even than the presence of armed guards in terminals is the time taken for and the detail of passenger and baggage screening. Passenger questioning, in attempts to determine if baggage might have been tampered with, if the passenger is carrying a package for anyone, and so forth is increasingly routine, even for U.S. airlines.

The simultaneous attacks in Rome and Vienna in December 1985 vividly demonstrated the vulnerability of airport terminals. Preliminary screening points at terminal entrances and restriction of terminal access to passengers and employees are intensive security options. Facilitated by the multiterminal layout of Kennedy Airport in New York, El Al Airlines currently employs this approach there.

OUTLOOK

The prospects are not encouraging. There is heightened sensitivity to the possibility of future attacks in traditional areas of international terrorist activity. The anticipated response of U.S. airlines operating in those areas is one of maintaining and intensifying extraordinary measures already in effect as necessary. The reaction of U.S. airlines operating domestically is more difficult to gauge.

It remains unlikely that organized terrorists could or would attempt to sustain a high level of violence in the United States within the next five to ten years, but the critical point is that it would not take many such spectacular acts to affect the public's perception of relative security and to generate a massive response to that threat. Depending upon the severity, the need for additional private security personnel, related services, and equipment would be substantial.

As trends in international terrorism have indicated, the airport terminal emerges as a particularly attractive and vulnerable target for terrorism within the United States. Although obviously not the only target of possible terrorist attack, it does present less risk than most to the terrorist and would be difficult to protect without major disruption to the operational efficiency of the commercial air transportation system. If attacked on a repetitive basis, a massive dilemma would be created for industry and governmental authorities. Under severe threat conditions, a decision to control access to terminals, not only departure areas as at present, would have tremendous economic and social implications, of the magnitude associated with the implementation of the total passenger screening program in 1973.

Technological Security

By FELIX POMERANZ

ABSTRACT: Losses from computer fraud and abuse are likely to have been substantial, and they continue to pose enormous potential risk. Nevertheless, technological advances have equipped the white hats with numerous effective countermeasures. Their response has continued to be diffused, however, because corporate executives do not believe that a significant risk exists, auditors have been loath to accept responsibility, and law enforcement officers are often averse to change. The answers to the problem lie in management awareness and in the managers' role as moral pacesetters within their organizations. The audit profession, currently in the throes of cataclysmic change, will adapt to public expectations, and information will be freely exchanged by the computer-using law enforcement officers. Universities will research perpetrators' motivations, their modus operandi, and the manner in which they can be detected—with the assistance of computerized data bases.

Felix Pomeranz is Distinguished Lecturer and director of the Center for Accounting, Auditing, and Tax Studies at Florida International University, and he is the Spear, Safer, Harmon Faculty Research Fellow. He has authored or coauthored four books and more than 100 articles, many of which are at the leading edge of audit technology. His books include Managing Capital Budget Projects: A Preemptive Audit Approach *(1984). He holds an M.S. from Columbia University, C.P.A. certificates from four states, and the designation of certified systems professional.*

THE potential loss from malicious or negligent computer use is immense. Further, most computer systems are undergoing continual change. These changes increase the systems' vulnerability to fraud because security measures tend to lag behind new technology.[1]

The nature and extent of the damages made possible by breakdowns in security can be only dimly perceived. Georgette Bennett, in her recent book *Crimewarps: The Future of Crime in America*, estimates the annual take of persons involved in street crime at $4 billion and that of "boardroom" and retail types at ten to fifty times that amount.[2] Similarly, the U.S. Chamber of Commerce reports that 30 percent of business failures each year result from the embezzlements, pilfering, and swindles of trusted employees. One expert has stated that three-quarters of all workers become thieves within five years of their employment and that up to 8 percent steal in volume.[3]

Many executives are unwilling to accept the possibility of computer fraud, although information specialists realize that the cost of security is far outweighed by the cost of mishandled computer information. One writer has urged executives to ask themselves what would happen if the accounts payable file disappeared, if information on research and development projects were turned over to a competitor, if inventory records were falsified, if customer orders were lost or stolen, if subsidiaries or divisions falsified information, if the production control schedule were modified, or if tax information were destroyed.[4] Put succinctly, the function of managing information systems may be the most valuable asset of an organization.[5]

Experts agree that top management must take the initiative if effective preventive measures are to be instituted.[6] Conversely, the head of the Internal Affairs Division of a large Florida bank has said that, in every single case investigated by his division, management involvement or lack of management awareness helped to make the crime possible!

THE DRAMA AND THE CAST

Dramatic changes in computer affordability and technology have led to widespread household ownership of computers and to international access to them through telephone lines. Contemporaneously, hackers have developed techniques that permit free access to phone lines and, therefore, to computer systems connected to these lines. The ability of one person to attack a system is thus multiplied by the others on the network. The potential for abuse and fraud is real, though overall costs of damage have yet to be determined.[7]

The list of hazards to which computers are exposed includes fire and physical damage, internal and external vandalism and sabotage, system invasion, interruptions of telecommunications, on- and

1. Candy S. Gilpin, "Electronic Cheating: How Safe Is the Computer," *Rough Notes*, 125(4):18, 34-38 (Apr. 1982).
2. Georgette Bennett, *Crimewarps: The Future of Crime in America* (Garden City, NY: Doubleday, Anchor Press, 1987), p. 104.
3. Ibid., p. 105.
4. Elizabeth Adams, "Secure or Insure," *Managing*, 1982, no. 1, pp. 21-26.
5. P. R. Hackenburg and J. E. Dugan, "Establishing Risk Parameters to Protect Computer Facilities," *Risk Management*, 31(11):20-30 (Nov. 1984).
6. Henry M. Kluepfel, "Computer Security: Shift into High Gear," *Security Management*, 30(9):116-21 (Sept. 1986).
7. Ian Murphy, "Aspects of Hacker Crime: High-Technology Tomfoolery or Theft?" *Information Age* (UK), 8(2):69-73 (Apr. 1986).

off-site power interruptions, facility security breakdowns, maintenance and machinery breakdowns, and loss of use and consequential damages.[8]

The list of perils is not becoming shorter. A recent article considers a new but seemingly old hazard. James C. Maxwell, a nineteenth-century mathematician and physicist, developed a theory that oscillating electrical charges generate electrostatic and magnetic forces able to penetrate most barriers. This theory spawned a new industry designed to protect information from compromising emissions. In 1982, a Dutch researcher showed that a cathode-ray tube display's high-frequency emanations could be picked up in television bandwidths and reformed as video signals; the extent of risk is indeterminate, because the phenomenon remains a point of contention in the industry.[9]

Bennett has made insightful comments concerning the "computer catastrophe":

Fortunately, large mainframe installations are compatible with security devices that can protect... programming... and help detect unauthorized entries. But, the 1 million minicomputer and 10 million personal computer systems on which small and medium-sized businesses rely, are much more vulnerable—and they outnumber the large systems by factors of two-to-one and twenty-to-one respectively.[10]

The situation has been aggravated by the decentralization of terminals and displays, in most cases unguided by headquarters' policies or personnel.

Law enforcement appears to have lagged in computerization. Bennett states that only two-thirds of police departments are using computer equipment to aid personnel deployment, long-term planning, and crime analysis; only one-tenth of the departments are said to use computers for fingerprint searches, cataloging of perpetrators' techniques, organizing investigations, and tracking offenders. Interchange of information between police departments has been complicated by numerous differing approaches to the formatting of data records and files.[11] The typical absence of information on either the modus operandi of the perpetrators or the manner of detection of the crime represents a crucial deficiency that should be redressed as quickly as possible.

Of course, in addition to police departments, the forces arrayed against malefactors include, first and foremost, management, especially security management, and internal and external auditors, private security services, and regulatory agencies. The public has long believed that auditors are responsible for the detection of fraud and, indeed, that this activity is the primary purpose for their existence. Contrarily, it has historically been the position of most external and some internal auditors that fraud detection is not of great importance and that professional responsibilities are circumscribed by considerations of materiality. The auditor's view is undergoing cataclysmic change, however, propelled in part by the growth of white-collar crime.

TRENDS IN TECHNOLOGY

Security measures, based on appropriate technology, should be incorporated into layers surrounding the asset to be protected. The layers are (1) physical countermeasures, (2) administrative

---

8. Hackenburg and Dugan, "Establishing Risk Parameters."

9. Vin McClellan, "CRT Spying: A Threat to Corporate Security," *PC Week*, 10 Mar. 1987, p. 35.

10. Bennett, *Crimewarps*, p. 108.

11. Ibid., p. 240.

countermeasures, (3) personnel countermeasures, and (4) computer-technology countermeasures.[12] State-of-the-art protective techniques can be grouped accordingly. What follows are series of security measures, many involving advanced computer technology.

Physical countermeasures

Physical countermeasures include the electronic surveillance of people and objects. This new surveillance technology differs from traditional social control in that (1) the technology is not impeded by distance, darkness, or physical barriers; (2) records are provided for easy storage, retrieval, and analysis; (3) the concern is with reducing risk and uncertainty; and (4) those under surveillance often become active partners in their own monitoring.[13] Comprehensive records of improved quality have become available, facilitating careful and in-depth study by corporate internal affairs personnel. In a somewhat different vein, it has become possible to create integrated building management systems that monitor and control life safety, intrusion, and environmental systems over a single cable, thereby simplifying wiring requirements. This capability resulted from multiplexing, the sequential or simultaneous transmission of various pieces of information at a very high speed over a common communications link.[14] It may also be more difficult to evade or neutralize the intrusion protection wiring.

Safes and vaults and bullet-resistant barriers are also among the physical countermeasures used. Sandia National Laboratories has had a major physical security program in effect for over ten years. The program includes identification of items subject to loss, assessment of risk, and selection of protective techniques that are of optimum effectiveness. Activities have ranged from component development and evaluation to full-scale system design and implementation. Sandia is engaged in efforts to transfer its technology to industry.[15]

Locks, passwords, access codes, and access cards are also physical countermeasures. Small cards may use different technologies, including magnetic stripe, magnetic dot, embedded wire, or passive proximity. Each technology has advantages and disadvantages and should be selected on the basis of an informed definition of what the access-control system is to accomplish. The effectiveness of the cards depends on the protection they are given by holders and on how difficult they are to forge or alter. Additional questions may arise as to whether secondary verification techniques should be used, given that such techniques may lengthen the access process considerably.[16]

Finally, physical countermeasures include biometric security techniques. These allow computer users to be identified by voice, fingerprint, hand geometry, or retinal patterns. Security is enhanced because access is allowed based not on what the person knows but on who the person is. For example, the Ridge Reader from Fingermatrix works

12. Kluepfel, "Computer Security."
13. Gary T. Marx, "The New Surveillance," *Technology Review*, 88(4):43-48 (May-June 1985).
14. Michael R. Tennefoss, "Alarm Monitoring with Time Division Multiplexers," *Security Management*, 29(2):37-43 (Feb. 1985).

15. D. L. Caskey, *Designing and Implementing Integrated High Security Physical Protection Systems: The Sandia Experience* (Washington, DC: Department of Energy, 1986).
16. Stuart Knott, "The ABCs of Access Control," *Security Management*, 31(5):84-89 (May 1987).

by scanning a person's fingerprint and using proprietary algorithms to compare the configurations of specific points along the print's ridges with those stored in the system for that person.[17]

Administrative countermeasures

Most administrative countermeasures are designed to separate the item to be protected from the potential perpetrator and to separate both of these from the knowledge required for access. Administrative measures against the misuse of computers include risk management approaches, which will be discussed later in this article; internal control standards, including audit trails; backup and storage policies and disaster recovery plans; insurance coverage; internal and external auditing; and access to security consultants.

Personnel countermeasures

Although this article focuses on technology, security is by no means an entirely technological issue. The human factors affecting data security should be reflected in a system's design and management. For a data processing operation, security design should minimize the number of people to be placed in a position of trust, as well as minimize the trust that should be placed in them. It is in the interest of the computer industry in general to establish a security ethic and to develop resources to raise the security awareness of the general public.[18]

Personnel countermeasures include

17. Sam Diamond, "Biometric Security: What You Are and Not What You Know," *High Technology*, 7(2):54-55 (Feb. 1987).
18. Vanagalur S. Alagar, "A Human Approach to the Technological Challenges in Data Security," *Computers and Security* (Netherlands), 5(4):328-35 (Dec. 1986).

comprehensive background investigations, polygraph tests, and voice-stress analyses. The techniques for a comprehensive background investigation are well known: reference checks, background inquiries, tests for honesty or for inclination to fraud, polygraph tests, and checks with bonding companies. Some industrial groups have established data bases of persons who previously have been convicted of or have confessed to a crime. When such persons reapply for a position in their familiar line of work, a search of the data bases is conducted—identification of felons through such searches has at times approached 1 percent. The effectiveness of all these screening techniques lies in the manner of their application.

A man whom I will call "Wilbur" sought a position as an internal auditor. In so doing, Wilbur assumed the identity of a foreign-born audit executive. Wilbur's falsification was detected by a bonding company investigator who discussed Wilbur's physical characteristics with persons given as references by the foreign-born auditor. The most recent communication pertaining to Wilbur was an inquiry from the warden of Fort Leavenworth, who wanted to place his newly arrived charge in suitable rehabilitative employment.

Computer-technology countermeasures

Countermeasures effected through computer technology include machine shutdowns after a number of unsuccessful access attempts and machine searches of a table of authorized transactions or of authorized users and of the extent of their authority. Authorization from a person with specific clearance may be required before a shutdown machine is restarted or processing resumes. Raychem has developed its own network management software program and a global communications network. The program ensures that each type of transaction is appropriately authorized by checking it against the firm's main data bases.[19]

Another tactic related to computers is encryption. Encryption uses a ciphering process similar to authentication in which a secret key makes the result unique for the user. When data are transmitted over radio-frequency links of microwave towers, they are available to anyone who can intercept them. To solve this problem, Hughes Communications Carrier Services chose Cylink's encryption hardware and software to secure its data transmissions.[20]

Another countermeasure is source-data automation. For example, firms are abandoning mechanical time clocks in favor of electronic systems that are comparable in price but eliminate labor costs and manual errors. Computerized units provide management with an effective means of integrating labor reporting into the payroll system and of monitoring employee activity. Electronic time clocks eliminate labor-intensive tasks, prevent employees from punching in too early or too late, eliminate underpayments or overpayments, and forward data to a payroll processor.[21]

Some specialized technology serving as countermeasures is related to credit cards and includes three-dimensional holograms, ultraviolet ink, fine-line printing, two-sided embossing, carbonless sales slips, and so-called smart cards. These techniques make it more difficult to duplicate or alter cards. Casio Microcard, among others, has developed a smart card that embodies an integrated-circuit chip that can increase card security and reduce fraud.[22] Another countermeasure comes from the Associated Credit Bureau, which keeps profiles of the characteristics of fraudulent credit-card applications.

In the area of retail sales, electronic tags, point-of-sales scanners, and television monitors are used. A unit of Onsite Research uses time-lapse cameras to monitor in-store traffic patterns and shoppers' behaviors. The system can (1) catalog interaction between consumer and product; (2) chart the selection process; (3) evaluate point of purchase displays; (4) examine traffic patterns; (5) evaluate employee training; and (6) serve as part of an employee-training program.[23]

Management's challenge is to select from the technological tools those that will be the most efficient and the most effective.

THE CORPORATE RESPONSE

To address computer risk successfully, management must first be committed to protecting the company's information, especially its proprietary information and trade secrets.

Risk management

Companies' protective actions, known collectively as risk management, are predicated on the theory that, to some

---

19. Ron Kopeck, "Raychem: Global Needs, Home-Grown Answers," *PC Week*, 10 Feb. 1987, p. 40.
20. Ron Kopeck, "T1 Encryption Plan Protects Data," *PC Week*, 3 Mar. 1987, p. 9.
21. Stan Gleich, "Electronic Time Clocks Are Effective Monitors," *Office*, 104(1):28 (July 1986).
22. "How Smart Cards Can Outwit the Credit Crooks," *Business Week*, 15 Oct. 1984, pp. 105-16.
23. "Taking the High Tech Approach to Pilferage," *Chain Store Executive*, 61(2):55-58 (Feb. 1985).

extent at least, unfavorable possibilities can be anticipated and their effects mitigated. Some risks can be avoided entirely by careful planning. Those that cannot be avoided must be assumed—which is normally done unwillingly except where the potential loss is deemed insignificant—or shifted to another, possibly an insurer. Even when a company succeeds in shifting the risk, it still has an important stake in minimizing its losses.[24]

The key tasks are to determine the risk parameters, to develop a plan of action, to implement that plan, and to reevaluate it continually in order to assure its viability.[25] The basic steps are as follows:

— list the potential causes of loss;
— estimate the value of the asset or information to be protected;
— identify the expected threats likely to cause loss;
— for each threat, estimate the probability of occurrence;
— quantify the estimated loss;
— determine the effectiveness of each technique that is to be used to protect the item;
— introduce the necessary safeguards.[26]

One commentator, writing in a banking industry publication, advised bankers to focus on insurance alternatives involving legal liability, disaster recovery, and computer security. Liability coverage for computer theft can protect vendors from intentional, unlawful employee acts that produce financial loss to customers; banks that provide data processing services were advised that they should consider purchasing insurance for data processing errors and omissions as a backup to the service contract.[27]

Preemptive approaches

Risk management is a sterling example of a preemptive approach, which incorporates the idea that management should improve control over operations before, rather than after, money has been lost. The audit counterpart, known as preemptive auditing, involves the evaluation of business plans, actions, decisions, and controls before actions are completed or decisions implemented.[28] Security functions represent a tried and proven preemptive management and audit application; indeed, security should be built into the management information system when it is first installed.[29]

Management must also create companywide awareness of the need for security. If that awareness permeates all levels of the organization, employees will introduce commonsense precautions likely to frustrate potential perpetrators. For example, modems will be switched off when not in use, thus precluding unauthorized access; passwords will be secured away from prying eyes; employment contracts will specify rights to intellectual property; copying machines will be monitored on a surprise basis; and so forth.

---

24. Felix Pomeranz, *Managing Capital Budget Projects: A Preemptive Audit Approach* (New York: Ronald Press, John Wiley, 1984), p. 29.

25. Hackenburg and Dugan, "Establishing Risk Parameters."

26. J. Fitzgerald, "EDP Risk Analysis for Contingency Planning," *EDP Audit, Control and Security Newsletter*, Sept. 1978, pp. 1-8.

27. Marr T. Haak, "The Bank Need for DP Insurance," *United States Banker*, 95(7):35-36 (July 1984).

28. Pomeranz, *Managing Capital Budget Projects*, p. 3.

29. Stephen Tarnoff, "Modern Criminals Prey on Computer Data," *Business Insurance*, 25 July 1983, pp. 23-24.

Segregation of duties

The ground rules for protecting assets or information are simple. In order for any abuse to occur, three elements must be present: first, there must be an asset or an item of information that can be converted or damaged; second, there must be a perpetrator with the personality makeup that inclines him or her to action; third, the perpetrator must have the knowledge to create an opportunity for access to the item. Thus control approaches have long been designed to separate potential perpetrators from assets and from the knowledge needed for access. Modern technology can contribute significantly to the achievement and maintenance of that separation.

With reference to computer fraud, studies have shown the need for foreclosing access opportunities by separation of custody of assets—or other items subject to conversion—from record keeping, of authorization of transactions from execution, and of planning from operations:

Segregation of duties is an environmental factor that must exist for controls over program development, program changes, and access to data files to function. Duties should be segregated within the data processing department and between the data processing and user departments.

The objective of segregation of duties is to have different people responsible for record keeping, physical custody of assets and general supervision and authorization of transactions. The data processing department should be organized to segregate responsibility for recording transactions and handling assets. The department should provide for separation of duties among three basic functions—operations, application development and maintenance, and data control.[30]

30. Ernst & Whinney, *Computer Fraud: A Report Presented to the National Commission on*

Periodic checkups

Periodic diagnostic checkups should be conducted by qualified consultants to ensure the continuing viability of the risk management program, to foster separation of duties, to confirm that loss prevention controls continue to function as planned, and to make certain that the company is in a position to take advantage of technological developments as they break. The idea is to keep the system ever responsive to changing needs. Stated differently, those who would breach security operate in a dynamic fashion—business cannot afford static systems.

The consultants' review should also include contingency plans, fire prevention and detection, water damage prevention and detection, prevention and detection of damage from air-conditioning failure, protection against other environmental threats, protection against hostile acts, and backup practices.[31]

Analytical procedures

Financial and operational information available on-line may be used to perform analytical procedures. Analytical procedures must be applied with in-depth knowledge of the company and its business. The results of these procedures should provide clues for investigating relationships that depart from informed expectations. For example, executives of a steamship firm, acting in collusion with stevedores, engaged in payroll padding and, eventually, in the unloading of fictitious vessels. This fraud was not

---

Fraudulent Financial Reporting (Washington, DC: National Commission on Fraudulent Financial Reporting, 1987), p. 14.

31. Pomeranz, *Managing Capital Budget Projects*, pp. 165-69.

detected by analytical procedures, however. The culprits had handled American short tons—2000 pounds per ton—but, unbeknown to reviewers, they reported British long tons, which each comprise 2240 pounds.

Many insurance companies have established in-house fraud units. The first line of defense is a computer that crosschecks claims and develops answers to leading questions such as:

1. Has the same person made several claims?
2. Has the same item been subject to multiple claims?
3. Does the claimant hold similar insurance policies with several companies?
4. Is the policy recently written?
5. Are the claimants in an accident all being treated by the same doctor?
6. Is the item claimed very expensive compared to the face amount of the policy?
7. Did a reported burglary take place while the claimant was on vacation?
8. Would a person really have lost a sable coat on a plane bound for Hawaii?
9. Can a person be suffering from an injured back if he or she was seen playing two hours of tennis?
10. Was business bad for the two years prior to the fire that destroyed the claimant's restaurant?[32]

To recapitulate, the keys to successful corporate countermeasures include top management's support, risk management techniques, periodic diagnostic checkups, and analytical procedures.

THE PROFESSIONAL RESPONSE

The white hats are starting to respond.

32. Bennett, *Crimewarps*, p. 157.

The auditors

The audit community consists of independent public accountants, internal auditors, and government auditors. The current climate within the audit community can be characterized as one of dynamic change.

In the words of the National Commission on Fraudulent Financial Reporting, "Generally Accepted Auditing Standards ... should restate the [independent public accountant's] responsibility ... to take affirmative steps to assess the potential for fraudulent financial reporting and to design tests to provide reasonable assurance of detection."[33]

At the time of this writing, the national commission's report, just quoted, and contemporaneous exposure drafts of the American Institute of Certified Public Accountants' Auditing Standards Board were undergoing evaluation by the accountants' congressional critics. It is hoped that the final rules will benefit the profession and its diverse publics. Technological developments will probably enable auditors to carry out any expanded charge. To give but one example, public data bases containing vast libraries of information have emerged; these data bases make it possible to execute a variety of innovative auditing steps. One group of such steps, concerning sensitization to potential difficulties, would include the following:

1. Identify significant legal actions—civil, criminal, or regulatory—to which the client, or his or her executives, may have been subject.
2. Inquire into information that may point to insolvency; examples include

33. National Commission on Fraudulent Financial Reporting, *Draft Report* (New York: American Institute of Certified Public Accountants, Apr. 1987), pp. 8-9.

negative trends, default on loan agreements, arrearages in dividends, denial of credit from suppliers, noncompliance with statutory capital requirements, and so forth.

3. Review filings with regulatory agencies for the names of related parties and for other entities in which officers and directors have ownership or other interests.

4. Refer to financial publications for the identities of unfamiliar customers or for the identities of other parties to transactions of questionable merit.

The expectation that on-line data bases will be useful to auditors is based on the presumption that repeat offenders are fond of iterating their modus operandi. Charles Harper, who directs the Securities and Exchange Commission's South Florida office, says, "You tell me the 'M.O.,' and I'll give you the name of the culprit."

Second, auditors may use information available on-line to perform analytical procedures using external or internal information. For example, an external review might involve a comparison of the client's financial or operational performance to others in his or her industry, with significant variations subject to investigation.

Third, on-line data bases may enable the auditor to verify management's judgments independently, thereby heading off audit failures involving management or financial fraud. The importance of management judgments is reflected in a study by the Securities and Exchange Commission concerning surprise write-offs. These write-offs were classified as follows: asset impairments, plant closings and restructurings, write-downs and write-offs of investments, and write-downs and write-offs of goodwill.[34] Opportunities for independent auditor verification include the following:

1. With respect to accounts receivable, review the adequacy of allowance for doubtful accounts vis-à-vis others in the industry; review open sales commitments, in light of industry conditions, in a search for possible losses; evaluate customer credit-worthiness.

2. With respect to inventories, review the nature of items in light of market forecasts and economic conditions; assess carrying values for obsolete or slow-moving articles, considering market conditions, customer preferences, changes in selling prices, capacity in the industry, and changing technology.

3. Concerning property, plant, and equipment, compare the client's policies to others in the industry with respect to depreciation and amortization, carrying values, write-offs, and maintenance; determine whether the industry has experienced idle capacity, abandoned property, or property held for sale.

4. Regarding investments, develop data for appropriate valuation of restricted securities, nonmarketable securities, or other investments that may have suffered impaired values; with respect to joint ventures, check the treatment to the partner's books.

5. Concerning goodwill, identify dispositions of major parts of the business to identify possible declines in value.

6. With respect to contingencies, develop estimates in light of economic conditions, settled litigation, and government regulations; check for open regulatory actions, fines, or assessments.

Further observations seem appropriate. To begin with, technology enables members of far-flung audit teams to be in constant communication with each

---

34. Dov Fried et al., "Surprise Writeoffs— Financial Reporting, Disclosure, and Analysis," cited in National Commission on Fraudulent Financial Reporting, *Draft Report*, p. 107.

other. Thus suspicious findings can be discussed on a real-time basis, and audit programs can be revised immediately. Also, as noted, auditing is more likely to become increasingly preemptive. Furthermore, it is likely that the auditor of the future will be required to attest to the system and its controls, as well as to the data processed; the precursor of this development can be seen in some public accountants' reports on computer software. These circumstances will combine to foster technological advances in auditing.

The lawyers

The legal profession, too, is being affected by technological developments, with effects paralleling those on auditors. Almost every law firm has some form of specialization in computer law. Lawyers engaged in this form of practice list the following as concerns: hacking, patent disputes, privacy, warranties, and product liabilities.[35]

The new technology has generated an array of related sociolegal issues. An indication of the problems may be gleaned by reference to the following matters, which appear to require clarification.

Ambiguous laws relating to computer crime. Many hackers who gain access to computer information without authorization will avoid criminal charges because of legal technicalities. The difficulty in prosecuting these cases stems from the many conflicting theft and computer-trespassing laws throughout individual states.[36] On the national level, the Counterfeit Access Device and Computer Fraud and Abuse Act of 1985 prescribes punishments for crimes involving federal computers, but there are defects or gaps in federal legislation overall.

Privacy. Given the geometric increase in types of electronic dossiers and given the modest protective legislation on the books, the climate for protection of information privacy does not appear favorable in the United States. First, there is no clear definition of what constitutes personal information. Second, many types of information are not covered by statutory protection. Third, no constitutional procedures exist for guaranteeing the accuracy and integrity of stored information.[37]

Electronic surveillance and civil liberty. The Office of Technology Assessment has prepared a report that, inter alia, deals with current and prospective use by federal agencies of surveillance technologies and with the interaction of technology and public law in the area of electronic surveillance. Special attention has been given to the balancing of civil liberties and investigative interests. The report includes suggested amendments to existing public law to eliminate gaps and ambiguities in current legal protection.[38]

Technology transfer. The Defense Advanced Research Projects Agency is concerned with technology transfer. All of the agency's contractors are also security conscious, but few, if any, are able to propose viable relief measures. There is a general fear that new restric-

---

35. James Connolly, "Patent Disputes, Hacking Major DP Law Issues in '85," *Computer World*, 21 Jan. 1985, pp. 14-15.
36. Ibid.
37. Gerard Salton, "A Progress Report on Information Privacy and Data Security," *Journal of the ASIS*, 31(2):75-83 (Mar. 1980).
38. U.S., Congress, Office of Technology Assessment, *Federal Government Information Technology: Electronic Surveillance and Civil Liberties*, 1985.

tions on communication within the agency's network could reduce creativity and productivity and might thus have a net negative effect on the nation's standing as a technology leader.[39]

In a similar vein, the Information Industry Association told the government that the new category of "sensitive, but not classified" information, and the use of the category, are confusing, threaten freedom of speech, and may have chilling effects on research, technological development, and business processes.[40]

Encryption standards. The Office of Technology Assessment invited computer industry vendors and consultants to discuss the power of the National Security Agency to establish encryption standards in both public and private sectors. Little consensus was achieved; rather, there was an expectation that the Congress might challenge the administration on security issues.[41]

The police

Technology is also starting to have a more pervasive effect on the nation's police departments. As Bennett writes:

Effective [police] resource management means maximum prevention, detection, and apprehension with minimum danger. . . . It also means that . . . cops must be savvy in the ways of microcircuits and sophisticated accounting.

39. Ronald G. Havelock and David S. Bushnell, "Technology Transfer at the Defense Advanced Research Projects Agency: A Diagnostic Analysis" (Manuscript, Technology Transfer Study Center, George Mason University, 1985).
40. Bill Dooley, "'Unsecrets' Data Curbs Worry IIA," *MIS Week*, 23 Feb. 1987, p. 1.
41. Vin McClellan, "Data Security Policy under the Spotlight," *Digital Review*, 12 Jan. 1987, p. 78.

In the future, police resources will be more mechanical and electronic than human. Computer-assisted information processing, electronic surveillance, advanced lie detectors, biochemical forensic tests, and nonlethal weapons will alter the methodology of law enforcement.

Our most innovative crime-fighting strategies are predicated on free-flowing intelligence. . . . It takes information to connect geographically remote crimes. . . . The FBI's magnificent psychological profiling program . . . would not [work] without information.[42]

CLOSING THOUGHTS

The potential loss from malicious or negligent computer use is immense. Further, numerous innovative technology-based techniques have been developed, many with positive implications for security. Nevertheless, the response of the good guys to the security problem has been slow and diffused. There has been a lack of awareness in many corporations, accompanied by an absence of risk management. Until just recently, auditors did not accept responsibility for the detection of even material fraud.

The data processing industry should adopt a security ethic and build greater security into its products. Universities should stress research into the origins and causes of fraudulent acts, based on an analysis of techniques of perpetration and methods of detection; a comprehensive national data base incorporating these concerns represents one important aspect of such an effort.

In the final analysis, however, the problem will be solved only by those managements that conduct business according to the golden rule and that set an example for the members of their organizations.

42. Bennett, *Crimewarps*, pp. 238-39.

Personnel Selection in the Private Security Industry: More than a Résumé

By IRA A. LIPMAN

ABSTRACT: Private security is charged with protecting both assets and people—employees, customers, vendors, even passers-by. This responsibility cannot be met without security officers who are honest, trustworthy, law-abiding, and psychologically stable. To guarantee a superior security force, the most stringent screening methods are required. The responsible private security company will use the best means of ensuring that its security guards, as well as all its employees, are of the highest quality. Those means include a combination of background checks, polygraph examinations, checks of criminal records, and drug tests. In addition, a psychological test should be administered to ensure that the person hired is not placed in the wrong job. Recent trends—social, demographic, and legislative—are making the selection process increasingly difficult. As a result, the quality of service provided by the industry as a whole will likely decline in the near future.

Ira A. Lipman is the founder and president of Guardsmark, Inc., the sixth largest security service company in the United States. He is chairman of the executive committee of the National Council on Crime and Delinquency. He received an honorary LL.D. from John Marshall University in 1970. He is the author of How to Protect Yourself from Crime *and was praised in a* New York Times *editorial for his leadership in the movement to disarm private security officers.*

THE issue of what constitutes permissible personnel selection methods is critical to the private security industry. With more than 1.1 million security officers protecting the assets of American institutions and the welfare of employees and nonemployees alike,[1] the issue is also critical to American society in general.

Incidents of security guards' committing crimes are legion, although studies do not seem to be available to determine if security personnel are actually more felonious than meat cutters or truck drivers. The nation's press certainly shows an interest in crimes committed by guards. Perhaps it is the irony of guards' violating what they are paid to protect that attracts the attention. Whatever the case, it is disturbing that so many who are hired to protect property either destroy it or steal it; that those especially trusted betray that trust; that those employed to exercise prudence are so often psychologically unbalanced.

The private security industry has a pervasive, if usually unnoticed, effect on public life-styles. We encounter private security personnel many times throughout the day and many more times than we see police officers. Apartment buildings, grocery stores, airports, sports stadia, office buildings, department stores, neighborhoods, factories, colleges, and courtrooms all rely on private security. The public should be able to rely on the honesty, probity, and professionalism of the private security officers who touch their lives at so many points. Unfortunately, that is too often not the case. Far too many security guards are unprofessional, dishonest, or psychotic, and far too many of these same dangerous individuals are armed.

A depiction of the private security officer as a psychotic who is armed and on the lookout for someone upon whom to exercise his or her authority would, however, be a caricature of the industry. Most security officers are hardworking, dedicated, careful individuals who take great pride in their chosen profession. Characterizing the average security officer as "an aging white male who is poorly educated and poorly paid"[2] is misleading; it is rather like saying that the man with one foot in a block of ice and the other in boiling water is quite comfortable on the average.

The true picture is quite different from the stereotype of the doddering pensioner guarding the local public library. Many security officers are college educated, highly professional, and well trained in security procedures. They handle volatile or potentially volatile situations with finesse and intelligence; they are trained in cardiopulmonary resuscitation, basic fire-fighting techniques, and loss-prevention measures; and they use highly sophisticated security equipment.

The responsibility of security officers to protect both assets and persons requires the use of the strictest personnel-screening methods available to private security firms. All of the screening methods are controversial to a greater or lesser extent. It must be remembered that, in the case of any particular position that is open, there are no fewer than four

---

1. William C. Cunningham and Todd H. Taylor, *Private Security and Police in America* (Portland, OR: Chancellor Press, 1985), p. 113. This study is commonly referred to as *The Hallcrest Report*.

2. James S. Kakalik and Sorrel Wildhorn, *The Private Police Industry: Its Nature and Extent*, vol. 2 (Santa Monica, CA: The Rand Corporation, 1972), p. 67. Guardsmark, Inc., is atypical of the industry in many ways, but the following statistics may be enlightening. Of its guard force, 10 percent have college degrees and 25 percent have attended college; 20 percent are female, and over 42 percent are minorities. The average age is 34.

interested parties: the applicant, the private security firm, the company purchasing the security firm's services, and that company's customers and employees. The applicant deserves a fair opportunity based on his or her record and character. The private security firm needs to hire the best available candidate. The company whose assets and employees the prospective guard will protect has the right to expect honesty, prudence, and professionalism. The employees and customers of, or visitors to, that company's facilities have the right to be protected or at least not harmed by the guard. The rights of an applicant must, of course, be weighed fairly, but they are counterbalanced by the rights of industry and the public.

SCREENING METHODS

All screening methods involve privacy issues as well as questions of fairness and reliability. A balance must be maintained between the rights of the applicant and the needs of the private security industry.

Background checks

The most basic selection method involves comparing the claims made by an applicant to the reality as pieced together from references and records of employment, education, and military service. It is surprising how many applicants lie on their application forms, and even more surprising that many private security firms fail to conduct more than a cursory background investigation. It is understandable that applicants strive to present themselves in the most flattering light, but outright prevarication is a different matter and is relatively easy to uncover. For example, it takes no great investigative wizardry to call a university to confirm that an applicant has actually earned the degree as he or she asserts.

Other major claims are just as easy to verify, including military service and type of discharge; previous employers, titles, and dates of employment; and awards. These facts represent only the bare bones of a character, however. To put flesh on this skeleton, to create a view of the whole person, personal evaluations are required.

References can be severely limited as sources of information for two main reasons. First, the references are selected by the applicant. They are more likely than not to paint a glowing picture of the applicant or they would not have been given as references. The second, and increasingly important, reason is the tendency of previous employers, because of potential legal liability, to reveal nothing about the applicant's job performance, whether good, bad, or indifferent. A good investigation will go beyond the personnel clerk to interviews with former supervisors and fellow employees who are more likely to yield solid information. The best source is often a former supervisor who is no longer with the company.

A detailed background investigation will also include a check of consumer credit reporting agencies and interviews with neighbors. An applicant who is financially irresponsible or heavily in debt would be a bad risk for a position as security guard. An applicant who is involved in altercations in his or her neighborhood might be likely to allow a disturbance at a protected facility to escalate out of control.

Polygraph tests

Space limitations do not permit an extended analysis of the controversy surrounding the use of the polygraph. The private security industry finds it essential, as does the president of the United States, the Department of Defense, the Central Intelligence Agency, the Federal Bureau of Investigation (FBI), the National Security Agency, and police departments across the country.

Critics of the polygraph test claim that it is not scientific, that it is not foolproof, and that interpretation of its results is a matter of subjective judgment. But that is precisely the nature of personnel decision making. However much personnel specialists try to apply rigor to their evaluations, the ultimate decision to choose one person over another is a subjective one.

The polygraph is a diagnostic tool and not a panacea. The results of a polygraphic examination must be weighed along with all the other information gathered about an applicant. Hiring decisions should never be made solely on the basis of what the polygraph indicates. If an applicant evinces physiological reactions to certain questions—for example, questions about previous thefts or drug use—those reactions are signals to the skilled polygrapher and personnel manager to probe further. The applicant might have something to hide, or he or she simply might have a tender conscience and an overactive imagination. Some people feel no guilt no matter how vicious a crime they have committed, while others feel guilt if they have only thought of committing even a minor crime.

The polygraph is also an excellent tool for verifying information that has been volunteered on the employment application. For various reasons, the background investigation might have failed to find support for an applicant's claims, particularly with regard to previous job performance and responsibilities. Former supervisors simply might not be available for comment, or they might adamantly refuse to depart from the company policy of providing nothing beyond dates of employment and title.

The polygraph can help a personnel manager decide if an applicant is lying about previous employment or if he or she is simply enhancing in a common and generally acceptable manner the responsibilities of a previous job. A former mail room clerk who says he was responsible for internal communications might not be lying but merely following the advice of some book on how to write a résumé.

Another important quality of the polygraph resides in its value as a deterrent. This works in two ways. Many potential applicants who know they could not pass a polygraphic examination given by a private security firm seek employment elsewhere. Deterrence also affects new hires, who are impressed with the employer's strict quality standards and willingness to incur extra expense to maintain those standards.

Drug testing

Because the topic of drug testing is covered elsewhere in this volume, it can be passed over here with just a note on the use of drug testing in the selection process. More and more companies are testing all applicants for drug use, a trend that will not subside in the foreseeable future. The need for such testing is clear, given the pervasive spread of illegal drugs in American society.

The claim has recently been advanced with regard to the drug-testing controversy that what an employee does on his or her own time is none of the employer's business as long as job performance is not affected adversely. For example, if it cannot be shown that using cocaine on Friday night lowers the user's efficiency and productivity on the following Monday, then the employer has no legitimate interest in the user's personal drug habits.

If that argument has any validity when applied to other industries, it certainly has none for private security. Willingness to break the law and risk addiction, job loss, reputation, and family welfare for the sake of a temporary sense of euphoria is something less than an ideal profile for a security officer. Perhaps there is no correlation between the quality of private life and the maintenance of high professional standards in some professions, but there must be in security. The Jekyll-Hyde type of character who can uphold the law during the week and break it only during off-hours is not the kind of person we want guarding our nuclear installations, hospitals, banks, and pharmaceutical companies.

A troubling aspect of drug testing is that job applicants will be tested on an entirely different basis from that for current employees. Few companies will try to dismiss an employee who fails a drug test—that is, tests positive for the presence of drugs—without confirming the results with a more sophisticated test. Employees have legal recourse that is not available to applicants, and companies will want to ensure a tight case so as not to leave themselves vulnerable to lawsuits. Applicants, however, lack any redress. Undoubtedly, an applicant will never know that he or she failed the drug test. Because false positives can result from the use of aspirin substitutes, antibiotics, diet pills, nasal decongestants, cold medicines, and prescriptions containing codeine, many applicants will be unfairly deprived of work opportunities.

Psychological testing

The profile of an applicant or new hire cannot be considered complete without the results of a psychological test, such as the Minnesota Multiphasic Personality Inventory (MMPI). Widely accepted and used, the MMPI consists of 566 questions that are designed to measure facets of personality. The test is generally acknowledged as the most objective means of assessing psychopathology.

The MMPI is particularly useful for security service applications, where determining risk factors is essential. An antisocial or aggressive personality type might be a fine candidate for *The Dirty Dozen* but not for a private security officer. An untold number of vicious and violent crimes are committed every year by guards, both armed and unarmed, who would never have passed the MMPI had they been required to take it.

The MMPI cannot currently be used for personnel selection, but it can be very useful in evaluating new hires for placement purposes.

Criminal records

It is not known how many security officers currently employed have criminal records. Judging from the number of cases in which a security guard commits a crime and is only subsequently found to have a criminal record, there are a good many. It is easy to understand why criminals would seek out positions in the security field. It is more difficult, however, to see how some security firms could be so irresponsible as to hire applicants without conducting a thorough check for past criminal activity.

The need for vigilance is constant. In the early 1970s, the owner of a security service firm informed the Pennsylvania legislature that he had to let go approximately 10 percent of his new hires after criminal records were discovered.[3] A decade later, New York State checked over 33,000 applicants' fingerprint cards for private security firms and found that around 10 percent had either an arrest or a conviction record.[4] The New York Investigation Commission found that two-thirds of the security guards working in the state in 1980 had arrest records.[5]

It might be thought that checking a person's criminal record would be an easy matter. Unfortunately, that is not the case. The system is inefficient and bureaucratic, and the check is time-consuming.

An effective criminal record check must be conducted on both the state and the national levels. In a typical search, a request is first made of the state, which checks its records and then sends the request on to the FBI. The FBI checks its own records as well as those of states that have sent their records to the FBI's Identification Division. The FBI requires the search to include fingerprint identification. The fingerprints are matched against the federal files and the state files that have been forwarded to the FBI. Any information uncovered at the FBI is then routed back to the state agency that initiated the search. The state agency in turn informs the private security firm of whether the applicant should be disqualified from holding a security position.

The FBI search can take two or three weeks, and the states vary greatly in their response time, which can extend up to six months. During this long process, the applicant has often become a provisional security officer—trained, uniformed, perhaps armed, and assigned to a facility.

This brief summary of the process does not convey the complexities and variations that exist. There is little uniformity among the states with regard to policies, procedures, costs, and length of delay. That convicted felons can be trusted security officers for even a month before their records are discovered is a situation that is both absurd and fraught with danger.

Criminal justice agencies conduct their own on-line searches via the National Crime Information Center telecommunications lines. In addition, twenty states participate in the Interstate Identification Index (III), a "national automated criminal history record exchange system."[6] The III is basically a data base that contains personal identification information but does not contain full criminal records. Instead, the system indicates where those records are located. The lack of uniformity across state statutes regarding access to criminal records by private security firms, however, effectively prevents state agencies from using the III for such searches.

Nevertheless, there is hope for access to the III. A proposal made in 1986 by SEARCH Group, Inc., a nonprofit national consortium for justice information, would give responsible state agencies access to the III for non-criminal-justice purposes, such as licensing and employment. States would no longer need to duplicate their files for transmis-

---

3. Milton Lipson, *On Guard: The Business of Private Security* (New York: New York Times Books, Quadrangle, 1975), p. 87.

4. *Security Letter*, 1 Nov. 1984.

5. Selwyn Raab, "Growing Security-Guard Industry under Scrutiny," *New York Times*, 4 June 1984.

6. "Proposed National Policy for Utilizing the Interstate Identification Index for Access to Criminal History Records for Noncriminal Justice Purposes" (SEARCH Group, Inc., 1986).

sion to the Identification Division, and the centralized state files now there would be returned to the states.

The states would remain in control of their criminal history records, the dissemination of those records, and the kinds of records—convictions, arrests, arrests still pending disposition, and so on—that could be disseminated. Criminal justice inquiries would still receive priority over non-criminal-justice inquiries, but for private security firms, the proposed changes would promise enhanced efficiency and fewer delays in qualifying applicants for security positions.

With all these methods[7] of personnel screening at the disposal of private security firms, why are so many security guards unstable, dangerous, and proven criminals?

A primary reason is that "state government regulation of private security can best be described as haphazard, fragmented and of little value."[8] As *The Hallcrest Report*, a major study of the industry, gently states, "The legislative provisions... do not appear stringent."[9] As of 1983, only 9 states required a fingerprint check of state criminal records and 5 required a check of FBI records for an applicant seeking a license to operate a private security firm. A copy of an applicant's criminal record was required by 15 states. In 29 states, a successful applicant could not have a felony conviction on his or her record. Some minimum level of training was required by only 16 states.[10] Many states that require criminal-record checks of owners of private security firms have no such requirement concerning guards.

The lack of stringent regulation and licensing requirements creates not only ease of entry for new companies but also an inclination on the part of companies to cut corners on personnel selection costs. The industry has traditionally been characterized by low margins, and most government contracts must by law be awarded to the lowest bidder. Control of expenses is therefore a necessity. The screening procedures previously described, however, though necessary for any private security company that demands excellence, are expensive. The total cost of screening one applicant can run into hundreds of dollars.

Low margins in the industry also necessitate generally low wage rates, particularly for companies that pursue low-bid contracts. The lower the wage rate, the lower the educational level and experience of the labor pool from which that employer can draw. Marginal personnel, little screening, and minimal or no training are a combination that is not conducive to excellence in security services.

FUTURE TRENDS

While recent years have seen some improvement in state regulation and licensing requirements, two major trends will make raising the general quality of private security forces more difficult.

The first trend is demographic. The Bureau of Labor Statistics projects that the number of 16- to 24-year-olds who enter the labor force in 1995 will be 10

---

7. Other tools are also used for screening in the private security industry, notably pencil-and-paper honesty tests and voice-stress analysis. Neither has received widespread acceptance.

8. Arthur J. Bilek, John C. Klotter, and R. Keegan Federal, *Legal Aspects of Private Security* (Cincinnati, OH: Anderson, 1981), p. 37.

9. Cunningham and Taylor, *Private Security*, p. 84.

10. Charles Buikema and Frank Horvath, "Security Regulation: A State-by-State Update," *Security Management*, Jan. 1984, pp. 40-41.

percent less than the number in 1987. At the same time, the number of jobs will grow by more than 10 percent,[11] and labor-intensive industries will show the greatest growth.[12] As a result, private security firms will find it increasingly difficult to recruit the best-qualified candidates. In addition, the seller's market will lead to general wage rate increases as businesses vie to fill their empty positions. If companies that hire contract security services refuse to accept rate increases, private security firms will be forced to employ less-qualified people who cannot command higher wages.

Even as the quality of private security begins to suffer from the mismatch between demographics and job growth, other trends make the problem worse. The District of Columbia and 12 states had bans on using the polygraph for preemployment purposes as of August 1987, and the ban may have become national by the time this article is published. A bill has passed the U.S. House of Representatives that prohibits the use of the polygraph in the private sector, with the exceptions of the drug industry and the contract private security industry. A bill was introduced in the U.S. Senate that would prohibit the use of all polygraphs in the private sector except for specific polygraph exams; this bill would effectively prohibit preemployment screening. The fear of liability on the part of employers is making it increasingly difficult to obtain evaluations of applicants' past job performance. Privacy laws have already sealed off bank and medical records. In addition, proof of graduation from high school no longer provides assurance of literacy and other basic skills.

For the private security industry, the paramount need is for honest, stable, qualified security personnel who can be relied on to carry out their responsibilities in a professional and dedicated manner. To acquire such a work force, stringent screening methods are essential. Such a work force is important, as American business and society increasingly depend on private security for protection against and prevention of crime.

The continuing tension in American life between the rights of the individual and those of society, or between those of the applicant or employee and those of the employer, is evident in the matter of private security personnel selection. There is clearly a societal interest in ensuring that an individual's rights to privacy and fair employment practices are defended. New social and technological developments will continue to renew the tension between the interested parties. Protection of privacy will be challenged by the surge in cases of acquired immune deficiency syndrome[13] and by breakthroughs in genetics that are imminent.[14]

It could well be that a marked lowering of the general quality of private security in the near future will lead to greater public and legislative demands that private security firms be enabled, or even required, to use the most stringent screening methods available.

---

11. "Help Wanted: America Faces an Era of Worker Scarcity That May Last to the Year 2000," *Business Week*, 10 Aug. 1987, p. 49.

12. Ibid., p. 50.

13. See Lindsey Gruson, "Privacy of AIDS Patients: Fear Infringes on Sanctity," *New York Times*, 30 July 1987.

14. For discussion of some of the ethical quandaries posed by genetic breakthroughs, see Robert Bazell, "Gene of the Week," *New Republic*, 23 Mar. 1987, pp. 13-14; Sandra Blakeslee, "Genetic Discoveries Raise Painful Questions," *New York Times*, 21 Apr. 1987.

The Legal Liability of a Private Security Guard Company for the Criminal Acts of Third Parties: An Overview

By JONATHAN D. SCHILLER and GARY K. HARRIS

ABSTRACT: The circumstances in which a private security guard company will be held monetarily liable to victims of criminal acts by third parties are examined. The standard of care imposed upon private security companies by general legal principles throughout the United States is identified as well as those classes of persons to whom such a duty is owed. Several common-law principles that mitigate the imposition of any liability imposed upon private security firms are also noted.

Jonathan D. Schiller received his bachelor's and law degrees from Columbia University and served as a law clerk to the Honorable Charles R. Richey, U.S. judge for the District of Columbia. He is a partner in the law firm of Rogovin, Huge & Schiller in Washington, D.C., specializing in litigation.

Gary K. Harris earned a law degree at Columbia University and served as a law clerk to the Honorable Ruth Bader Ginsburg, U.S. circuit judge on the U.S. Court of Appeals for the District of Columbia Circuit. He practices law at the law firm of Rogovin, Huge & Schiller in Washington, D.C.

IT is, of course, wholly unremarkable to recognize that it is vital that American industry maintain adequate security against both internal and external unlawfulness. Losses from armed robbery, burglary, arson, vandalism, shoplifting, industrial and foreign espionage, and employee theft cannot only cripple an individual enterprise[1] but hamper the national economy as well.[2] Losses in productivity resulting from illegal drug use and assaults by fellow employees or third parties are similarly staggering.[3]

It is somewhat more noteworthy that business losses from such acts are increasing dramatically and, according to the Department of Commerce, have reached "epidemic levels."[4] In response to this rapid growth, industry is increasing its use of private security guard companies to prevent these losses from occurring or at least to minimize them.

In addition to assisting in minimizing these losses, however, the introduction of private security guards into the workplace raises its own set of potential problems for American industry. Simply by increasing the number of persons with authorized access to business property, the potential for internal theft is concomitantly increased. Similarly, unauthorized or excessive acts by security guards in pursuit of their stated objective, crime prevention, can, through imposition of legal liability and disastrous public relations, detrimentally affect the businesses they are assigned to protect.[5]

Because a private security guard company is vicariously liable for such unauthorized or excessive acts by its employees,[6] the business should be able to recoup from its retained guard company all tangible losses suffered as the proximate result of the guard's unlawful acts.[7] Moreover, in light of valid social policies and a long history supporting the doc-

---

1. A 1983 study by the Department of Commerce concluded that approximately 30 percent of all business failures annually are the result of employee theft. See U.S., Congress, House, Subcommittee on Education and Labor, *Polygraphs in the Workplace: The Use of "Lie Detectors" in Hiring and Firing: Hearings on H.R. 1524 and H.R. 1924*, 99th Cong., 1st sess. 1985, p. 331 (hereinafter cited as *Polygraph Hearings*).

2. Employee theft alone, for example, costs business $40 billion per year. Retail consumer prices are 10-15 percent higher to cover such internal losses. See *Polygraph Hearings*, p. 305. Successful foreign espionage, naturally, injures U.S. national security as well as the economy.

3. According to the Alcohol, Drug Abuse and Mental Health Administration, lost productivity in 1986 resulting from alcohol and drug abuse was approximately $100 billion.

4. *Polygraph Hearings*, p. 331.

5. Such unauthorized or excessive acts by security guards are not limited to particular industries or to geographic regions. Rather, they occur across the board due to poor judgment in selecting the private security guard company, overdependence on armed security guards, or just bad luck. For example, in a Miami bank, a recently hired security guard, without warning or provocation, shot and killed a fellow guard. Although the killer was declared incompetent to stand trial and was committed to a state mental hospital, the victim's widow sued the bank, alleging that, because the bank knew, or should have known, of the killer's mental instability, her husband's death was the result of the bank's negligence in hiring the killer for armed guard duty. Rather than risk a jury award in excess of $1.5 million, the bank paid $300,000 to settle the suit. *Wall Street Journal*, 30 Apr. 1987, p. 25. An enterprise can, and should, attempt to minimize the potential for such instances by retaining only private security guard companies of the highest quality, ones that utilize the most advanced methods, such as psychological testing, in screening and assigning their employees. Similarly, the use of security guards who are armed should be kept to an absolute minimum.

6. William Prosser et al., *Prosser and Keeton on Torts*, 5th ed. (St. Paul, MN: West, 1984), pp. 505-6.

7. Due to a business' difficulty in proving the amount of damages it suffered as a result of any adverse public relations from such an incident,

trine of vicarious liability,[8] one could not argue that the private security guard company should escape legal liability in such circumstances.

There are, however, more difficult and troubling questions regarding the legal liability that may be, and often is, imposed upon a security guard company. For example, suppose that during an armed robbery of a convenience store to which one private security guard is assigned, a customer of the store is shot and seriously injured. Is the security guard company liable for all damages the customer suffers?

As with most, if not all, legal questions, the answer is simply, "It depends." There are various circumstances in which a private security guard company will be held liable for the criminal acts of third parties. First, the standard of care that the law imposes upon a private security guard company retained to provide security to a particular establishment must be explored. Then the circumstances in which the law imposes monetary liability on a person or enterprise for the criminal acts of others must be examined. As unfortunate as it may be, a security guard company will often be held liable for the criminal acts of third parties.

THE STANDARD OF CARE IMPOSED UPON A PRIVATE SECURITY GUARD COMPANY

It is generally well settled[9] that a business owes a nondelegable duty to its employees,[10] contractors, employees of its contractors, and customers to provide a reasonably safe environment, considering the nature of the premises, or to warn them of the risks of unsafe conditions that the enterprise should realize others may not reasonably discover.[11] This standard of reasonable care is defined to require the enterprise to conduct its operations in light of "such knowledge of the conditions likely to harm [others] as persons experienced in the business and having special acquaintance with the subject matter have."[12]

Where an enterprise retains a private security guard company to fulfill its legal obligation to make the workplace reasonably safe, the private security company assumes that legal obligation and is therefore held to the same standard of care as is the contracting party.[13] Thus the private security guard company must exercise reasonable care in performing its duties.[14]

---

such losses may go unrecompensed. See generally Dan Dobbs, *Remedies* (St. Paul, MN: West, 1973), pp. 150-57.

8. Prosser et al., *Prosser and Keeton on Torts*, p. 500.

9. Because the law with regard to the issues discussed herein may vary from state to state, this article must confine its scope to the principles set forth in the series *Restatement of the Law*, published by the American Law Institute, which presents the generally accepted majority position with regard to the various issues it addresses.

10. A nondelegable duty may be delegated to another for performance; but if the person to whom performance of the duty is delegated acts improperly, the enterprise remains subject to liability to his employees. *Restatement (Second) of Agency* (Philadelphia: American Law Institute, 1958), chap. 4, topic 4, title C, introductory note, p. 435 (hereinafter cited as *Agency*).

11. Ibid., section 492.

12. Ibid., section 495.

13. *Restatement (Second) of Torts* (Philadelphia: American Law Institute, 1965), section 324A (hereinafter cited as *Torts*). This doctrine, in which one agrees, either for consideration or gratuitously, to render services to another that "he should recognize as necessary for the protection of a third person," is sometimes referred to as the Good Samaritan Rule. See, for example, *Rick v. R.L.C. Corp.*, No. 82-1059, slip op. at 4 n.1 (6th Cir. 1983).

14. *Torts*, section 324A.

Because the private security guard company, however, is retained for its expertise in protecting against crime and thereby ensuring a safer environment, the standard of care to which it will be held is significantly higher than that to which its client would be held.[15] For example, a client business's duty to provide a safe environment includes a duty to inspect the business, though the extent and frequency of inspection depend upon the nature of the things to be inspected, the danger to be anticipated if inspections are not made, the ability to make inspections without seriously interrupting the business, and all other factors involved in the determination of the reasonableness of conduct.[16]

Because a security guard company's sole function may be to ensure a safer environment, the security guard company may be required to conduct more frequent and more thorough inspections of the premises than the law would require of the business itself.

Moreover, the persons to whom a private security guard company owes this duty of reasonable care extend far beyond the enterprise that retained it and with whom the guard company entered a contract. Rather, the private security guard company owes this duty to all persons who would reasonably expect to be protected in the circumstances, such as employees, customers, contractors, and other invited guests of the business the guard company is hired to protect.[17]

The private security guard company, however, is not held to be an insurer. Rather, the guard company will be held liable only if its malfeasance was the proximate cause of the injury suffered.[18] Thus, for example, even though an employee or customer may be seriously injured by the criminal acts of a third party, the private security guard company will not be liable to the injured party unless it failed to perform its tasks with reasonable care.

The rigors of the common-law duties imposed upon the guard company are also mitigated by the doctrines of contributory negligence and assumption of risk. In common law, a victim's contributory negligence is a defense to an action for harm suffered by a violation of another's duty of care.[19] This rule bars recovery by a victim, even when hurt by a conscious failure to maintain safe conditions as long as that failure was not intentional or wanton.[20] The effect of the contributory-negligence doctrine, however, is limited by the last-clear-chance rule, which permits a contributorily negligent victim to recover from a defendant who had notice of the victim's perilous position and who, through reasonable care, could have

15. See *Agency*, section 495; see also text accompanying footnote 12 of this article.
16. *Agency*, section 503, comment c.
17. *Torts*, section 344. This principle is a significant departure from the nineteenth-century rule that an agent, such as a guard company, owed a duty only to its principal, the person or enterprise on whose behalf the agent acted. See *Agency*, section 355, comment f.

18. *Agency*, section 354 and comments a, b; see also *Agency*, section 352. Additionally, for the security guard company to be held liable to such third persons, the injury suffered by the victim must be physical, rather than economic, loss. See *Agency*, section 357.
19. *Torts*, section 467. Many jurisdictions, however, have now adopted a comparative negligence approach, in which a victim's own negligence is not necessarily a complete defense. Rather, it may simply permit a pro rata apportionment of damages. See generally Prosser et al., *Prosser and Keeton on Torts*, section 67.
20. See generally *Agency*, section 525, comment a.

prevented harm from coming to the victim.[21]

The assumption-of-risk doctrine is another defense to a victim's legal action for breach of the private security guard company's duty to ensure a safe environment or to warn others of the known dangers.[22] As set forth in the *Restatement (Second) of Torts*, the assumption-of-risk doctrine is as follows:

[A person] who fully understands a risk of harm to himself... caused by the defendant's conduct... and who nevertheless voluntarily chooses to enter or remain... with the area of that risk, under circumstances that manifest his willingness to accept it, is not entitled to recover for harm within that risk.[23]

Thus even "if the private security guard company negligently or intentionally fails to perform what would otherwise be its duty, a [victim] who becomes aware of a dangerous condition . . . ordinarily has no cause of action for harm thereby suffered."[24] There are, however, several exceptions to the assumption-of-risk doctrine. One merits particular attention with regard to the instant issue: where an actor "manifests his intention to remedy dangerous conditions of employment" but fails to do so, the assumption-of-risk doctrine does not apply.[25]

In light of these well-settled legal principles, it is clear that a private security guard company is required to exercise reasonable care in performing the duties it has been retained to provide. Moreover, its reasonable care must reflect its expertise and experience in providing such services. A failure to meet this reasonable-care standard will be deemed by the law to be negligence.

Will a security guard company's negligence therefore subject it to monetary liability to a victim of a third party's criminal act? At first glance, it would appear that no one, other than the criminal, should be held so liable, but the common law, reflecting centuries of experience and attempting to implement society's values, is often not what a first-glance conclusion would indicate. It is to that issue that we now turn.

LIABILITY FOR THE CRIMINAL ACTS OF A THIRD PARTY

To prevail in a negligence action against a private security guard company, a victim must establish not only that the guard company was negligent but also that the security guard company's negligence was a proximate cause of the injury or injuries sustained.[26] In defining what constitutes proximate or legal cause of the harm sustained by a

---

21. *Torts*, sections 479-80; see also *Agency*, section 525, comment a.
22. *Torts*, section 496A.
23. Ibid., section 496C(1).
24. See *Agency*, section 497, comment f.
25. See *Agency*, section 522. Moreover, there are some circumstances in which the security guard company's duty is unqualified. If it becomes known to the guard company that an employee of the business that retained the guard company, while acting within the scope of this employment, either has come "into a position of imminent danger of serious harm" or has been hurt and become helpless, then the guard company has a duty in the first instance "to exercise reasonable care to avert the threatened harm" and in the second instance to use reasonable care to give first aid and care for the employee until he can be cared for by others. *Torts*, section 314B.
26. "In order that a negligent actor shall be liable for another's harm, it is necessary not only that the actor's conduct be negligent toward the other, but also that the negligence of the actor be a legal cause of the other's harm." *Torts*, section 430.

particular victim, the *Restatement (Second) of Torts* states as follows:

The actor's negligent conduct is a legal cause of harm to another if
(a) his conduct is a substantial factor in bringing about the harm, and
(b) there is no rule of law relieving the actor from liability because of the manner in which his negligence has resulted in the harm.[27]

Where harm results from the intervening acts of third parties—as would always be the case in the situations currently under review—causality will generally be a contested issue.

The general rule regarding the effect of intervening actions by third parties is that "an actor may be liable if his negligence is a substantial factor in causing an injury, and he is not relieved of liability because of the intervening act of a third person if such act was reasonably foreseeable at the time of his negligent conduct."[28] This rule applies even where the intervening conduct is criminal in nature.[29] This modern rule of law, imposing liability where the risk of a criminal act is foreseeable, stands in sharp contrast to "well-established principles that . . . a criminal act of a third person is generally a superceding cause" of the victim's harm.[30] Thus, in circumstances where a specific criminal act was not reasonably foreseeable, a private security guard company may find some support for the argument that any lack of care on its part was not the proximate cause of, for instance, terrorist-related injuries.[31]

The law, however, appears to address directly the circumstances currently under review.

If the likelihood that a third person may act in a particular manner is the hazard or one of the hazards which makes the actor negligent, such an act whether innocent, negligent, intentionally tortious, or criminal does not prevent the actor from being liable for harm caused thereby.[32]

Because one of the prime motivations underlying the retention of a private security guard company is the prevention of crime and because that is one of a guard company's major responsibilities, it would appear that, under the standard previously quoted, the private security guard company will be held liable if a violent criminal act was attributable to, in part, its negligence. It would appear that only a very unusual criminal act would be deemed unforeseeable in circumstances in which a private security guard company is present. Indeed, it is typically only because criminal acts by outside third parties are foreseeable that the private security guard company has been retained.

The court's decision in *Meyser* v. *American Building Maintenance, Inc.* provides an example of the application

---

27. *Torts*, section 431.
28. *Vesely* v. *Sager*, 5 Cal.3d 153, 163 (1971); see also generally *Torts*, section 442A.
29. *Torts*, section 448.
30. See generally *Torts*, section 440.
31. See, for example, *7735 Hollywood Blvd. Venture* v. *Superior Court*, 172 Cal. Rptr. 528, 529

(Ct. App. 1981); see also generally *Torts*, section 448. In *7735 Hollywood Blvd. Venture*, for example, the court held that a landlord's failure to replace an outdoor light bulb did not breach any duty to a tenant raped by one who broke into the building. In so doing, the court observed, "In this day of an inordinate volume of criminal activity . . . no one really knows why people commit crime, hence no one really knows what is 'adequate' deterrence in any given situation." 172 Cal. Rptr., p. 530. The court went on to state that the fact that "anyone can foresee that a crime may be committed anywhere at any time" does not, "per se, impose a duty to install security devices meeting a jury's concept of adequacy." Ibid.
32. *Torts*, section 449.

of this legal principle.[33] In *Meyser*, a laundromat was robbed and damaged by arson set to cover the robbery. Thereafter, the owners of the laundromat sued the firm it had retained to provide security for the laundromat, claiming that the robbery and arson were attributable to the security firm's negligence. The jury unanimously found in favor of the owners of the laundromat. In affirming the jury's verdict, the court of appeals stated, "Under the circumstances such as those presented there, where defendant was hired for security purposes, foreseeability of arson or other invasive criminal acts is apparent." The court of appeals' strong language in *Meyser* confirms, as noted earlier, that it will be unlikely that a court will accept the argument from a private security guard company that a criminal act was unforeseeable in the circumstances.

CONCLUSION

A private security guard company must exercise reasonable care under the existing circumstances. Because of the high proficiency and expertise implied with respect to a reasonable private security firm, the reasonable-care standard imposed upon such enterprises is concomitantly high. Additionally, the law holds that this duty of reasonable care is owed by the private security company to wide classes of persons, including customers and employees of the business by which it was retained.

The law has also determined that, because the threat of criminal acts by third parties is precisely one category of hazards that the private security company has been retained to prevent, such criminal acts are foreseeable. Due to their foreseeability, the law imposes monetary liability upon a private security guard company whose failure to meet the standard of reasonable care was a factor in permitting the criminal act to take place. Because the company that retained the private security firm will also be held liable in such circumstances, prudence counsels that the business retain only private security firms that employ the most advanced techniques in personnel selection, assignment, and crime prevention.

33. 85 Cal.App.3d 933, 149 Cal. Rptr. 808 (1978).

The Time Has Come to Acknowledge Security as a Profession

By ERNEST J. CRISCUOLI, Jr.

ABSTRACT: Despite the importance of private security in today's world, security has yet to become widely accepted as the complex and demanding profession it really is. The successful practice of security requires extensive specialized knowledge as well as a full range of general managerial skills. The field has witnessed much progress in the last decade toward codification of its body of knowledge, development of formal academic programs, and establishment of professional certification programs. The two biggest obstacles to acceptance as a profession are erroneous public perceptions of the field and the lack of a structured prerequisite to practice. Both of these obstacles are being addressed by the industry's leading association and others in the field.

Ernest J. Criscuoli, Jr., has been executive vice-president of the American Society for Industrial Security (ASIS) since May 1977. A certified protection professional, Criscuoli spent 18 years with the General Electric Company, ending as security manager for General Electric's Valley Forge Space Center. He served 10 years on the ASIS board of directors and was ASIS's twentieth president and chairman of the board. Criscuoli also was security manager at the Curtiss Wright Corporation's Research Division and served in the U.S. Army Counterintelligence Corps. He is a graduate of Boston College in economics.

ALTHOUGH the roots of private security can be traced back many centuries—at least to the eleventh century, long preceding public law enforcement, which did not surface until 1783[1]—many people have never thought of private security as being a profession. Granted, security does not have a highly structured academic prerequisite to practice as many professions do; neither does it require a license or certification to practice, although guard personnel must be licensed in some jurisdictions. Yet, it bears remembering that all professions, including medicine and law, were practiced for centuries before the path of entry to them was made formal or eligibility to practice them was regulated.

What set certain fields apart from other livelihoods and marked them as being professions was the need for specialized knowledge. The formal definition of a "profession" is "a vocation or occupation requiring advanced training in some liberal art or science, and usually involving mental rather than manual work, as teaching, engineering, writing, etc., especially medicine, law or theology."[2] In light of this definition, security can be considered a profession because it requires "advanced training" of a "mental rather than manual" nature.

This claim is not made to suggest that all security practitioners demonstrate full professional competence; no profession could seriously make that claim. Rather, it is made to point out a fact many people seem unaware of—that security is not merely a matter of intuition or common sense; it involves a complex body of knowledge, analytical abilities, and the know-how to prescribe suitable security measures for individual circumstances, as well as the effective use of an array of other managerial skills.

TYPES OF KNOWLEDGE AND SKILL REQUIRED

Up-to-date familiarity with physical security devices and controls and their uses is one of the most obvious knowledge requirements for security professionals. What may not be so obvious is the breadth of that subject. By itself, physical security occupies many specialists full-time. Access controls, for example, range from simple locks to complex computerized systems, and the 1987 *Security Industry Buyers' Guide* contains more than 125 such categories of equipment. But what security professionals must know does not end with physical security.

They must also be aware of legal considerations and labor relations issues and how these should and do affect corporate security policies and practices. Legal requirements and government regulations pertinent to security in the particular industry in which the security practitioner's employer is engaged must be known and adhered to. The security professional must also be ever mindful of the potential for conflicts between the organization's need for security and individuals' rights to privacy.

Procedural knowledge is yet another requirement for security professionals. Security surveying, vulnerability assessment, risk analysis, personnel screening methods, subject matter and techniques for training, loss reporting and analysis, proper investigative techniques, contingency planning, and other procedures specific to security must be mastered.

In view of the scope of technical

---

1. Victor Green and Ray Farber, *Introduction to Security: Principles and Practices* (Los Angeles: Security World, 1975), p. 25.

2. *Webster's New Universal Unabridged Dictionary* (Simon & Schuster), 2d ed., s.v. "profession."

knowledge involved in security, Bennett Hartman has concluded that, like practitioners in law and medicine, security managers cannot possibly commit to memory all there is to know in their field. Therefore knowing broadly what information exists and where to look for specific information when needed is a key requirement for security managers. Staying abreast of scientific and procedural advances as well as legal developments must be continual, accomplished through review of the industry's literature, participation in educational programs, and interaction with colleagues in the field.[3]

Many security practitioners have gained much of their technical knowledge through experience in public law enforcement, the military, and the academic arena. At one time, mere possession of technical knowledge was sufficient for an individual to perform effectively as a security practitioner and to be considered a professional. In the last 10 to 15 years, however, more has been expected from the security practitioner than just technical skills.

Today, security requires a broad range of management expertise as well as knowledge of the technical aspects of security. Security practitioners are expected to be cognizant of the way the client's or employer's business is run and to be oriented to what contribution security can make to the overall success of that business. The security practitioner needs to be knowledgeable about and to understand such management concerns as return on investment, budget and finance, personnel matters, compensation, public relations, and insurance and liability issues.

3. Bennett Moyses Hartman, "The Need for Professionalism in Security Administration" (Ph.D. diss., Pacific Western University, 1978), p. 3.

Because costs for security programs have, like most business functions, escalated significantly over the past decade, security professionals must identify the most cost-effective means of providing protection. This means integrating personnel, equipment, and procedures in an efficient manner that causes the least possible interference to ongoing business operations while adequately safeguarding the organization's assets.

Hartman has pointed out that professionalism is vital to a "security program," given the "size and complexity of present day organizations. [Consequently, security personnel at all levels must] be well trained in their specific functions."[4]

Any security administrator who is unable to transcend the company-cop mentality and deal with business associates as a business manager will inevitably fail to provide fully effective security. Cooperation and support from management and employees as a whole are key facets of security programs. If employees feel the security administrator is out to get people rather than to assure a safe and secure working environment, that cooperation and support will not be forthcoming.

Unlike other business professions such as purchasing, finance, and personnel recruiting, in which the application of fundamentals is much the same from one organization to another, security differs considerably according to the specifics of the organization where it is applied. To be sure, there are similarities, and knowledge can be transferred, but how these are adapted to various situations is crucial to the effectiveness of the security effort.

For example, the measures used to

4. Ibid., p. 1.

secure a large plant involved in research and development differ greatly from those suitable for a chain of small retail outlets, computers, oil fields, transportation operations, high-rise office buildings, hotels, or communication networks. Yet, because the security practitioner may be called upon to provide protection for such a diverse collection of operations, the knowledge needed to match preventive measures to a given situation is also necessarily broad.

Security for one company may consist of protecting a headquarters facility, one location where access is relatively easy to control. In another organization, such as a nationwide chain of convenience stores, facilities must be protected in all kinds of locations—suburbs, the middle of big cities, low-crime areas, and high-crime areas—so the task is considerably more complex. The tools and methods that are effective in the first instance cannot be transferred directly to the other.

In some organizations, securing the manufactured product, such as television sets, is the focus of efforts, while in others, the security of the components that go into making the final product is equally important. Computer chips are an example that readily comes to mind. Valuable and easily concealed, they make attractive targets for internal theft. In still other organizations, ideas and information are critical assets, and even though they are intangible, the security professional must devise ways to safeguard them.

Because the products and services to be protected are as diverse as business itself, the security professional must be virtually a Nostradamus to project what might be at risk and what threats must be protected against. Threats may come from people within the organization, from people outside it, or even from natural disasters. The security professional must determine what is needed, in light of the potential for any given threat, to protect the organization's assets adequately. Whatever is selected must still permit a free flow of personnel, ideas, and materials for manufacturing, distribution, and service activities.

The 1980s have presented security practitioners with diverse challenges, from drug and alcohol abuse to industrial espionage and terrorism. In addition, while in the past, security was of concern primarily to defense contractors, in recent years, even the Boy Scouts of America and religious institutions have found it necessary to call on the security profession for help. Because so many more kinds of assets and businesses now require protection, many more solutions to the safeguarding problem must be devised.

WHY IS PROFESSIONALISM IMPORTANT IN SECURITY?

The role of security has taken on a much greater importance over the past two decades. Lives, as well as organizational survival, are often dependent upon the effectiveness of security. For this reason, competent performance by security professionals is critical.

Private security as we know it today originated as a Department of Defense requirement and was handled under other key organizational functions, including personnel, finance, and legal. As recently as twenty years ago, the security manager's only responsibility was the safeguarding of government classified information. Security was considered simply a cost of doing business with the government. So long as government inspections of the contractor's facility were satisfactory, nothing further was demanded. Many veteran security practitioners now refer to those as the good old days.

During the 1950s and early 1960s, security managers were rarely asked to provide total security for a given plant or industry. The social unrest of the late 1960s and the changes in the social climate in this country since that time, however, have thrust security managers into new arenas of responsibility. To a significant extent, this occurred because, at about this same time, public law enforcement was suddenly faced with providing more protection while undergoing reductions in personnel and funds. Much of the burden was shifted to the private sector, which had traditionally obtained some, if not all, of its protective services from local, state, and federal law enforcement.

The new reliance on the security profession also arose because, in the 1970s, companies that had no government-classified work, and therefore little previous interest in security, began to realize that personnel and property were subject to increasing risks that threatened the well-being of the company. Their rising need for protection could not be accommodated by public law enforcement, so private security practitioners were turned to. Personnel screening, risk management, executive protection, drug abuse prevention, and responses to a host of other problems that now confront American business became security responsibilities.

Security has played an expanded role for nearly two decades and is being recognized in more and more organizations as an essential element of organizational survival. Because top management calls more frequently upon the security director and can be expected to do so even more in the future, a trend to place security personnel closer to the chief executive officer in the organizational structure can be observed.

Despite this trend, security professionals are not being given carte blanche to accomplish their mission. Consequently, the security director must be able to show how sound security measures can enhance and contribute to business objectives, or else the resources necessary for adequate security may be limited. Showing the contribution a secure environment makes to productivity by relieving employee concerns about their safety relates the expenditure for, say, a computerized access-control system to a fundamental company objective and improves the likelihood of the acceptance of the expenditure.

WHY IS SECURITY NOT WIDELY ACCEPTED AS A PROFESSION?

A major obstacle for security professionals is the general public's failure or unwillingness to perceive security as a profession. People are ignorant of what security professionals do primarily because security practitioners have not actively publicized their roles in organizations. By the nature of their positions, they tend to work without great display and often behind the scenes.

Misconceptions have been perpetuated largely because the public's awareness of security most often results from contact with uniformed personnel and news reports of security gone awry. Just as a receptionist often conveys the initial impression of an organization to visitors, the uniformed officer is the first impression many people receive of the private security organization. Yet most activities for which security professionals are responsible are seldom seen by the general public and virtually never make the news.

Several efforts have been made in recent years to correct misconceptions about security and to raise awareness of what security professionals do. The American Society for Industrial Security (ASIS) recently prepared a short videotape entitled "The Invisible Man" to inform general business audiences of the breadth and complexity of the security field. Educating general management about security was deemed important partly because the lack of understanding often leads management to hire poorly qualified individuals to fill incorrectly defined jobs, with the result that the "business executive becomes dissatisfied with the security profession."[5]

The importance of the "basic problem" cannot be overestimated. Management's losing sight of the "need for professionalism [among] security personnel" creates the false impression that "any person with a law enforcement background can perform key security functions without further and specialized training. . . . In the past, billions of dollars have been lost through misjudgment in proper selection of personnel."[6]

In addition, organizations all too often pay not for professional security performance but for the learning experiences of individuals hired from fields viewed as related to security, even though those backgrounds actually offer little loss prevention experience. During the learning period, the employer's security programs may be weakened, and once the person has gained adequate knowledge, the employer frequently loses the individual to another organization. "The original employer has . . . financed the training, but has not [received any]

5. Ibid., p. 2.
6. Ibid., p. 28.

benefit from it—another expensive failure."[7]

This pattern of misinformed management's hiring of persons with only law enforcement, intelligence, or investigative background contributes to a primary source of misconceptions about the security profession—the influx of people to security as a second career who mistakenly assume there is little or no difference between what is done in public law enforcement and what is done in security. Although many individuals from public-agency backgrounds quickly make the transition, others do not, and those individuals convey an erroneous impression of what competent security is all about.

The private sector's goal is the prevention of crime, whereas law enforcement focuses primarily on investigating crimes after their occurrence and apprehending criminals. Of course, security does involve some investigation and, on rare occasion, apprehensions, and in some law enforcement agencies, specially assigned officers do concentrate on preventive efforts.

Another significant difference is that private industry is structured differently from a public law enforcement agency. The public sector spends taxpayers' money for the common good. By contrast, in the private sector, the goal is the production of products or services at the least cost possible in order to return the maximum to the stockholders or investors. Also, in the private sector, relatively few rules and regulations guide one's conduct, whereas in public law enforcement, most actions are taken as a result of laws, codes of conduct, and regulations. In that sense, law enforcement can be considered a more exact science than can private security.

7. Ibid.

WHAT HAS OCCURRED TO ADVANCE THE PROFESSION?

Even in the days when security was concentrated in the business of defense contractors, the need to share knowledge and upgrade capabilities was recognized. The security managers for those contractors formed ASIS for that purpose. As a result of the rising demand in other industries for individuals who could oversee programs to protect organizations' assets, ASIS's membership rolls swelled from a mere 6000 in 1975 to almost 25,000 in 1987 and the society's focus has broadened considerably. ASIS is now the world's largest professional membership organization serving security.

A network of more than 180 ASIS chapters fosters information exchanges on a local basis, and with security's expansion into all kinds of organizations, thirty committees are dedicated to particular specialties in the field, from banking and finance security to museum, library, and archive security, and from educational institutions security to telephone and telecommunications security. A number of relatively small associations have also cropped up that represent individual security specialties.

To assist its members with staying abreast of the many facets of the profession, ASIS publishes *Security Management* magazine. A reflection of the expansion of the industry, *Security Management* has increased from under 400 pages a year to more than 1300 pages a year. The field is also addressed by several commercially owned magazines and a variety of special-focus newsletters, two of which are devoted entirely to legal decisions relevant to security.

Much has occurred to advance the profession on the academic front as well. In the late 1970s, the reduction of funds for law enforcement resulted in a decrease in trainees and declining interest of students in law enforcement courses offered at academic institutions. The concurrent rise in private security led to the introduction of security courses at two- and four-year colleges. In 1970, the American Association of Junior Colleges indicated a total of two associate-degree programs in the United States offering security and loss prevention programs. By 1975, the number of institutions offering courses had risen to 113.[8]

As late as 1978, there were no master's degree programs offered in industrial security.[9] A few master's programs are available today, and in 1986, the ASIS Foundation joined forces with Central Michigan University to sponsor a program that will lead to a master of science in administration with emphasis in security.

While security was gaining a foothold in the academic community, ASIS began to offer workshops that stress the technical aspects of the profession. It also offered week-long courses in the Assets Protection series, which now includes Assets Protection Courses I, II, and III as well as the Professional Certification Review, a two-day course.

ASIS began work on a professional certification program in the early 1970s. "Those of us in the security field never really had any objective standards for evaluating professional competence," according to Timothy J. Walsh, CPP. The tenth president of ASIS, Walsh, a recognized leader of the security profession, was involved with the Certified Protection Professional (CPP) program from its inception. "We were looking for

8. Richard S. Post and Arthur A. Kingsbury, *Security Administration: An Introduction*, 3d ed. (Springfield, IL: Charles C Thomas, 1977), p. 771.

9. Hartman, "Need for Professionalism," p. 58.

a way professional colleagues could recognize some common level of achievement."[10]

The 1976 *Report of the Task Force on Private Security*, by the National Advisory Committee on Criminal Justice Standards and Goals, included "professional certification programs" as a specific goal that "can strengthen the role of private security personnel and increase the professionalism of the industry."[11]

A 1977 statement from ASIS's Professional Certification Board pointed out that

both the American Society for Industrial Security's decision to establish the professional certification program and the Task Force on Private Security's recommendation on developing voluntary certification programs for private security managerial personnel are based on the fact that establishment, maintenance, and development of standards are means that have been historically used by the professions and other fields requiring special education, knowledge and experience to assure the necessary quality of service to society by those who desire to practice in a particular profession or field.[12]

The Professional Certification Board's statement also explained that

the following points were determined as having principal validity in decisions to proceed with a certification program for protection professionals:

1. Other professional groups and individuals, as well as general management, increasingly depend on security and loss prevention principles and practices. Some of their direct and delegated responsibilities involve liabilities under laws and regulations, and they are increasingly concerned that others on which they depend are likewise responsible and measured against standards of excellence and performance.

2. From the government and public viewpoints, the primary issues are ones of public safety and welfare. In the same sense that other professionals' performances are monitored when these issues are involved, so increasingly is the performance of the security professional as the conduct of his duties impacts on public safety and welfare.

3. Since certification is not a function of a government agency, it seldom has legal status. However, in an increasing number of cases in recent years, certification has been granted government recognition as an indication of a person's professional competence.

4. From an employer's perspective, certification can provide a meaningful standard, in addition to information as to academic record, position experience, and professional performance, for evaluation of those desiring employment, promotion, and advancement.

5. Certification tends to improve professional competence in the field and thus promotes the public's welfare and raises the public's respect for the profession.[13]

The first CPP was designated in 1977. ASIS followed the excellent work by the American Society of Safety Engineers and many other organizations that were operating successful certification programs. In taking these steps, ASIS was the first to acknowledge formally that those who would fill the role of security director in the future would need certain specialized knowledge and skills and academic credentials to meet future security challenges.

According to Dick Cross, ASIS president in 1973 and another leader in development of the CPP program,

There were three goals behind the establishment of the program. First, ASIS had a public and organizational responsibility to

---

10. Shari M. Gallery, "Doctor, Lawyer, CPP," *Security Management*, 31(1):50 (Jan. 1987).
11. Hartman, "Need for Professionalism," p. 53.
12. Ibid., reprinted by permission.
13. Ibid., pp. 53-54, reprinted by permission.

provide some mechanism for security practitioners to demonstrate their professional ability. Second, we needed to elevate the personal status of security professionals in both government and industry, as they were not viewed very highly. And finally, employers in need of a security person would have some assurance of that person's level of knowledge.[14]

One of the primary aims of the CPP program has been to encourage members of the profession to keep up with change. Toward this end, recertification based on qualifying activity points is required every three years.[15]

Analyses of candidates for the protection professional certification have confirmed that formal schooling in traditional police or law enforcement subjects does not enable candidates to score substantially higher on the CPP exam than do majors in other subjects. In addition, while many candidates have law enforcement, intelligence, or investigative backgrounds, such experience has not, in itself, proven an adequate identifier of protection professionals.[16]

As professional certification was developing, the tools available to security practitioners changed and advanced markedly, benefiting from developments in the computer industry. Alarm systems, as one example, now operate by computers and integrate many functions other than signal intrusion, including selective access control, fire protection, environmental control, and closed-circuit television.

The tremendous growth in security equipment and services has caused the exhibits portion of ASIS's annual seminar to grow from 128 booths in 1977 to over 650 booths in 1987, making that event a major educational resource for the security manager. The new capabilities made possible by advancing technology necessitate a much greater effort to stay up to date. To aid security professionals further with locating the equipment and services best suited to specific assignments, ASIS assisted the publisher of the *Security Industry Buyers' Guide*, Bell Atlantic, in developing its format and content.

To ensure that the body of knowledge of the security profession is accessible, ASIS established the O. P. Norton Information Resource Center in 1985 at its headquarters in Arlington, Virginia. Currently staffed by two full-time professional librarians, the center's holdings are being computerized to provide ready access for security managers.

Still another sign of advancement in the profession can be seen in the ASIS Foundation, which raises funds to underwrite scientific research, expand educational pursuits in the field, and support scholarship programs for security students.

CONTINUING CHALLENGES FOR THE PROFESSION

As a profession, security is at the stage where finance and data processing were twenty or thirty years ago. Much progress has been made toward formal identification of the field's body of knowledge and establishment of academic programs addressing that subject matter. The CPP program has passed the 10-year mark and has gained wide acceptance as an indicator of professional knowledge. This accomplishment unquestionably marks a turning point in the struggle to establish security as a profession. Expansion on the progress on these fronts can be expected to strengthen security's position as a profession.

---

14. Gallery, "Doctor, Lawyer, CPP," p. 50.
15. Ibid., p. 58.
16. Ibid., p. 56.

Security continues, however, to face an uphill climb in its efforts to gain acceptance as a profession among the broader business community and the general populace in the United States. A concerted effort to acquaint persons outside the field with its scope and complexity is imperative. Until those who hire individuals to fill security positions understand the qualifications necessary for effective performance, the field will be vulnerable to the image problems caused by unqualified persons who are ineffective in their efforts. Unfortunately, the inabilities of individuals are all too often generalized to the profession as a whole.

Further, if the security job itself has not been appropriately conceived, even a fully qualified security professional may not be able to fulfill management's expectations. Therefore, the importance of expanding understanding of the profession among top management is doubly important, first so management can properly define the role of security in the organization and then so qualified individuals will be selected to fill security positions.

Educating entrants to the field as to the requirements for professional practice and how security differs from other fields they may have previously worked in is a further necessity. An especially important aspect of this education process is emphasis on the need for broad management skills and understanding of business goals and how security fits into them.

Finally, individual practitioners must demonstrate professional competence and communicate an accurate picture of the security profession to business colleagues if long-standing stereotypes are to be supplanted by the image of a professional manager whose technical expertise lies in security.

Can Police Services Be Privatized?

By PHILIP E. FIXLER, Jr., and ROBERT W. POOLE, Jr.

ABSTRACT: Some people consider police services as inappropriate for privatization, arguing that such services are public goods that only government can practically provide. The work of E. S. Savas and others, however, has persuasively demonstrated that many government services are not public goods or, at least, not pure public goods. Police services, in fact, have been successfully financed, through user fees, and delivered, via contracting, by the private sector. Moreover, there are some surprising examples of fully privatized police services, both financed and delivered privately. The major barriers to police privatization include tradition and attitudes, concern about control and accountability, union opposition, legal restrictions, and the difficulty of encouraging all beneficiaries to finance these services voluntarily, or privately. All of these barriers can be surmounted under certain circumstances. There are even signs that the privatization of police services, especially some milder forms of privatization, is gradually taking place.

Philip E. Fixler, Jr., is director of the Local Government Center, a research institute that studies privatization at the local and state government levels. Fixler also edits Privatization Watch, *the nation's principal monthly newsletter on privatization.*

Robert W. Poole, Jr., is president of the Reason Foundation, a free-market-oriented think tank. He was the first in the United States to promote the concept of privatization of government services. Poole cofounded the Local Government Center in 1976 as a think tank to research and publicize privatization.

ONE of the most significant developments in state and local government over the past decade has been the privatization revolution. Numerous public services have been shifted, in part or in whole, into the private sector.

Public safety and criminal justice functions have shared in this privatization trend. There are hundreds of for-profit emergency ambulance firms and several dozen private fire protection, jail, and prison operators. There has also been extensive use of private, voluntary alternatives to civil court proceedings, and a growing amount of contracting with private organizations for work-release programs, juvenile rehabilitation, and the like. Policing, however, has generally been considered a service that cannot and should not be privatized.

How coherent is this point of view? Are there characteristics of the police function that render it so inherently governmental that it ought not even be considered for privatization? And if this widespread view is mistaken, how do we account for its persistence among public policy analysts?

POLICING AS A PUBLIC SERVICE

When analysts refer to certain types of services as being inherently governmental, what they generally mean is that the services in question have the characteristics of public goods, as opposed to private goods. In simplest terms, a public good is one that is provided collectively and from whose benefits those not paying for the good cannot be excluded. The classic example of a public good is national defense. If an organization provides defense of a given territory, everyone within that territory receives the benefits, whether or not they pay the costs. Thus clearly identified public goods are generally produced in the public sector and paid for via taxation.

Privatization analyst E. S. Savas has developed a somewhat more sophisticated typology of public and private goods.[1] Savas uses two basic criteria—exclusion and joint consumption—to categorize various services. A good that is consumed privately and available only to those who pay for it is a private good. Savas points out, however, that some goods that are consumed collectively may still be charged for individually in proportion to use—for example, cable television, water and electricity service, and toll roads. Savas terms them "toll goods." By contrast, a public good is one that is consumed collectively and for which nonpayers cannot be excluded.

How should we categorize police services in this typology? The key to answering that question is to realize that policing is not a single service. Police departments perform a number of functions, some of which have the characteristics of private goods, some of which are toll goods, and some of which are, in varying degrees, collective, or public, goods. This being the case, it is clear that there is at least some theoretical scope for privatization in police services.

Some police department functions are essentially of a security guard nature, providing specific protective services to a specific client. Examples include surveillance of a vacant house when the owner is on vacation, police escorts to a funeral, or providing traffic direction at a construction site that blocks a lane of traffic. In Savas's typology, these are basically private goods, which could be provided by a private firm. Some police departments charge for such services.

1. E. S. Savas, *Privatizing the Public Sector* (Chatham, NJ: Chatham House, 1982), pp. 29-51.

Other aspects of security-related police work are less private in nature. If a neighborhood, a shopping center, or even a specific block receives regular police patrol, those who live and work there will feel safer than otherwise. In such cases, the consumption of the service is joint, and the key privatization question is whether funding can be provided voluntarily, rather than by taxation. This type of police service falls between a toll good and a collective good.

In addition to understanding the different types of services that police provide, we must also understand the various forms of privatization.[2] To do so, we need to look at two key dimensions of public service delivery: who pays for the service and who delivers it. The traditional form of public service—and the general assumption for all police functions—has the government providing the funding via taxes and directly producing the service using government employees, but private mechanisms may be used in either or both of these areas.

Thus, if government produces the service but charges individual users, in proportion to their use, the funding—but not the delivery—of the service has been privatized via user fees. On the other hand, if government retains the funding responsibility, collecting taxes to provide the funds, but hires the provider in the marketplace, we have the form of privatization known as contracting out. Finally, if both the funding mechanism and the service delivery are shifted into the private sector, we have the most complete form of privatization, referred to as service shedding or load shedding. Government may retain some degree of control over the terms and conditions of service, as in the case when it issues an exclusive franchise. Alternatively, government may simply bow out, other than for ordinary business licensing, leaving the service to be handled entirely by the marketplace.

EXPERIENCE WITH POLICE PRIVATIZATION

All three types of privatization have been experienced in the United States with respect to policing: user fees, contracting out, and some degree of service shedding. The following subsections provide an overview of these cases.

User-financed police services

While many jurisdictions have traditionally charged special fees for some types of services, such as providing security for parades and special events, one of the most common user fees is a charge for responding to burglar alarms. Oftentimes, these charges are for responding to false alarms, usually after a certain threshold point.

The growing number of home owners who have contracted for alarm services has placed an increasing burden on public police departments, even with the imposition of charges for responding to false burglar alarms. As a result, some police departments have permitted alarm companies to provide response services to answer their subscribers' alarms. When Amarillo, Texas, authorized a local security company to respond to subscribers' alarms, the police department was relieved of the time-consuming responsibility of answering an average of eight alarms per day; it thereby saved approximately 3420 man-hours, or the equivalent of adding one and three-fourths people per year. All of this was

---

2. See Charles Feinstein, *Privatization Possibilities among Pacific Island Countries* (Honolulu: East-West Center, 1986), p. 10.

at no cost to the taxpayer.[3]

Another, more comprehensive, form of user-financed police service is exemplified by the arrangement negotiated between the Montclair Plaza shopping center and the Montclair, California, Police Department. Calls for service from the shopping center became so numerous that the department considered designating the center as a new, separate beat. Because a limited budget precluded establishing such a beat, the department proposed that the shopping center pick up 50 percent of the beat officer's salary and benefits.[4] The arrangement proved to be mutually beneficial. The department was able to accommodate the new service demand, while the shopping center and its security staff achieved greater liaison with the department.

A similar concept was implemented in Oakland, California, where private developers entered into an agreement with the police department to fund special downtown patrol services to attract more shoppers. According to James K. Stewart, director of the National Institute of Justice, "Fear is down in the inner-city areas and development is thriving."[5]

Contracting out: Private provision of police services

As with a number of other public services, contracting out police services began with support services and later certain auxiliary services. Police departments in many parts of the nation contract out for vehicle maintenance and other support services. In some cities, the towing of illegally parked vehicles is still a police responsibility, but nearly three-fourths of all cities have privatized this function. New York City uses two different methods in privatizing it. First, New York contracts with private firms to tow these vehicles. In addition, the city charges those claiming their towed vehicles a fee to cover the police department's costs.[6]

Among the support services that have been contracted out are communications-system maintenance, police training, and laboratory services;[7] food provision and medical care for jail inmates;[8] and radio dispatching services.[9] Traffic control and parking are other police responsibilities that have been contracted out in some jurisdictions. Several large security companies provide special-event security, and other companies provide such mundane services as guarding school crossings. The city of Los Angeles, for example, has contracted out for school crossing guards. Private firms furnish parking lot enforcement services for the Eastern Idaho Medical Center, the Arizona Department of Transportation, and the University of Hawaii, Hilo.[10]

Contracting out line law enforcement

---

3. Dale Pancake, "Cooperation between Police Department and Private Security," *Police Chief*, June 1983, pp. 34-36.

4. Roger M. Moulton, "Police Contract Service for Shopping Mall Security," *Police Chief*, June 1983, pp. 43-44.

5. James Stewart, "Public Safety and Private Police," *Public Administration Review*, Nov. 1985, pp. 758-65.

6. John Diebold, *Making the Future Work* (New York: Simon & Schuster, 1984), p. 161.

7. Institute for Local Self-Government, *Alternatives to Traditional Public Safety Delivery Systems: Civilians in Public Safety Services* (Berkeley, CA: Institute for Local Self-Government, 1977), p. 14.

8. John J. McCarthy, "Contract Medical Care," *Corrections Magazine*, Apr. 1982, pp. 6-17.

9. Philip E. Fixler, Jr., "Which Police Services Can Be Privatized?" *Fiscal Watchdog*, Jan. 1986, no. 111, p. 3.

10. Ibid.

activities is an obvious second level of police services contracting. For example, San Diego, Los Angeles County, and Norwalk, in California, and St. Petersburg, in Florida, have contracted out some public-parks patrol.[11] Private security guards also provide protective services in Candlestick Park in San Francisco and the Giants Stadium in New Jersey.[12] Other jurisdictions—including San Diego; Lexington, Kentucky; and New York City—have contracted for private patrol of crime-ridden housing projects.[13] The Suffolk School District in New York contracts for school security services.[14] In Munich, Germany, private police patrol the subways.[15]

Many state and local governments in this country contract for public buildings and grounds security, among them being Boston; Denver and Fort Collins, Colorado; Houston; Los Angeles County; Pensacola, Florida; New York City; San Francisco; Seattle; and the states of California and Pennsylvania.[16] Many departments of the federal government also contract for private security guards, and U.S. federal courts in many parts of the nation use private court-security officers and bailiffs who are sworn in as U.S. deputy marshals.[17]

Overseas, the British Transport Police stationed at the Port of Southampton, who had been public employees since 1820, were recently replaced with private security personnel.[18] In several other countries, some licensed security personnel are authorized to exercise police-like powers.[19]

Prisoner custody is also being turned over to contractors by some jurisdictions. For many years, the transportation of prisoners has been contracted out in parts of Maryland and California.[20] Santa Barbara County, California, contracts for some of its prisoner transportation, for example. Private police have also been hired for other specialized custodial services, such as guarding prisoners who are being transported or who are being treated in public hospitals. For example, New York City and several Alabama prisons, including the West Jefferson, St. Clair, and Hamilton facilities, contract with private firms to guard prisoners receiving hospital treatment.[21]

Choosing to contract out for custodial care has moved beyond these special situations, however. At least six U.S. counties now contract with private organizations to manage local prisons; these countries include Hamilton County, Tennessee; Bay County, Florida; Butler County, Pennsylvania; Hennipin County, Minnesota; Santa Fe County, New Mex-

---

11. Robert W. Poole, Jr., *Cutting Back City Hall* (New York: Universe Books, 1980), p. 40.

12. Martin Tolchin, "Private Guards Taking the Place of Police," *Rutland Herald*, 29 Nov. 1985.

13. Poole, *Cutting Back City Hall*, p. 40.

14. Institute for Local Self-Government, *Alternatives to Public Safety*, p. 14.

15. Poole, *Cutting Back City Hall*, p. 41.

16. Ibid.; Fixler, "Which Police Services Can Be Privatized?" p. 2.

17. Wackenhut Corp., "Court Officers Now in 21 States," *Wackenhut Pipeline*, Dec. 1984, pp. 1, 4.

18. "Public Police Singing the Blues," *Reason*, Feb. 1985, p. 13.

19. Clifford D. Shearing and Philip C. Stenning, "Modern Private Security: Its Growth and Implications," in *Crime and Justice: An Annual Review of Research*, ed. Michael Tonny and Norval Morris (Chicago: University of Illinois Press, 1981), 3:193-245.

20. Philip E. Fixler, Jr., "Private Prisons Begin to Establish Track Record," *Fiscal Watchdog*, June 1986, no. 116, pp. 1-4.

21. Steve Joynt, "Private Companies Hired to Guard Hospitalized State Prisoners," *Birmingham Post-Herald*, 22 May 1986; Fixler, "Which Police Services Can Be Privatized?" p. 3.

ico; and Aroostock County, Maine. In 1986, California and Kentucky became the first two states to contract out the management and operation of state prison facilities. California's contract is for a minimum-security, return-to-custody facility for parole violators in Hidden Hills. Kentucky has gone a step further by contracting with a firm that owns its own site.[22]

There is even some precedent for contracting out a basic police service such as investigation. Some federal agencies contracted out for many investigative services up until about 1909.[23] About twenty years ago, the governor of Florida had a short-lived contract for investigative services, to be paid for from private contributions. Some police departments in the Midwest contract out for special narcotics enforcement.[24]

A third level of police privatization is contracting for regular or full police services in a given jurisdiction. The federal government has contracted for full police services at its Mercury test site, operated by the Energy Research and Development Administration in Nevada, and at the National Aeronautics and Space Administration's Kennedy Space Center in Florida.[25]

Perhaps most far reaching, however, are the several examples where local governments have contracted for regular police service and even their entire police force. Apparently the first in modern times to do this was the city of Kalamazoo, Michigan. In the mid-1950s, Kalamazoo contracted for about three and a half years with a local firm for street patrol and apprehension of traffic-law violators.[26] In order to ensure that the contractor's activities were fully in accordance with the law, patrol personnel were sworn in as full deputy sheriffs. A court case—involving a technicality relating to an arrest—led to the demise of the arrangement, even though the decision was in favor of the private firm. One of the dissenting judges made a virulent attack on the whole concept of private police, helping to discredit the practice in Kalamazoo.

In 1981, Reminderville, Ohio, contracted out its entire police service. Previously, Reminderville had received policing under contract from Summit County. When the county withdrew its patrol from Reminderville—and unincorporated Twinsburg—for budgetary reasons, Reminderville and Twinsburg first explored the possibility of re-contracting with the county at a higher rate. At that point, a private security firm, headed by a former Ohio police chief, submitted a competing bid of $90,000 per year, about half the county's bid. Moreover, the firm agreed to provide two patrol cars rather than one and to reduce emergency response time from 45 to 6 minutes. As part of the arrangement, the firm also agreed to "select trained, state-certified candidates for the police positions, leaving the village to make the final choice. In fact, village officials would have full autonomy in hiring, firing, disciplining, and organizing the police force."[27] The arrangement worked well for two years, surviving threats of a lawsuit by the Ohio Police Chief's Association. Nevertheless, an

---

22. *State Government News* (Commonwealth of Kentucky), 30 Dec. 1986.
23. George O'Toole, *The Private Sector* (New York: Norton, 1978), p. 28.
24. Ibid.
25. Poole, *Cutting Back City Hall*, pp. 41-42.

26. William Wooldridge, *Uncle Sam, the Monopoly Man* (New Rochelle, NY: Arlington House, 1970), p. 122.
27. Theodore Gage, "Cops, Inc.," *Reason*, Nov. 1982, pp. 23-28.

attack in *Newsweek* and other skeptical publicity disturbed Reminderville officials, and in 1983, they ended the contract and set up their own conventional city police department—at a higher cost.

A similar instance occurred several years earlier in the small town of Oro Valley, Arizona. In 1975, a local company offered the town full police services as part of a comprehensive public safety package including fire protection and ambulance service.[28] The company "agreed to establish a police headquarters and keep all records according to state guidelines for police departments" and to supervise and assume all liability for the conduct of its employees. Oro Valley's town marshal, however, would fully control and have responsibility for the police force and would be able to override the firm's authority at any time. The price for full police service was substantially lower than what it would have cost the town to set up its own public police force.

This time the contracting-out arrangement was undermined when the Arizona Law Enforcement Advisory Council refused to accept the firm's employees in its training and accreditation program—and the council is the only organization licensed to award accreditation.[29] This development was followed by a state attorney general's opinion that Oro Valley could not commission the firm's employees as police officers. In the face of anticipated high court costs, the company decided to discontinue the arrangement.

Several other small American towns have contracted for private police services for as long as five years, including Buffalo Creek, West Virginia; Indian River, Florida; and several other small Florida[30] and Illinois[31] jurisdictions. Sometimes these contracts were an interim measure until the jurisdiction could form its own police force.

The most recent example of contracting for police services was in connection with the Burbank-Glendale-Pasadena Airport Authority's contract with Lockheed Air Terminal for security services. Unfortunately, a flaw in the 1982 legislation specifically authorizing the hiring of private law enforcement agents with police powers was recently discovered and the arrangement was terminated.[32]

The only other country in which there has been any significant amount of contracting for regular police services is Switzerland. Some thirty Swiss villages and townships contract with a firm called Securitas. According to the Swiss Association of Towns and Townships, contracts offer substantial savings over what it would have cost these small towns to operate their own police forces. The typical contract calls for foot and vehicle patrol, checks on building security, night closing of bars and restaurants, and ticket validation at special events.[33]

Privatizing finance and provision: Deputization and special powers

The most extensive form of privatization ranges from situations where private security personnel are given arrest pow-

28. Ibid.
29. "Oro Valley Must Hire Own Police," *Arizona Territory*, 19 Feb. 1976.
30. Poole, *Cutting Back City Hall*.
31. William C. Cunningham and Todd H. Taylor, *Private Security and Police in America* (*The Hallcrest Report*) (Portland, OR: Chancellor Press, 1985), p. 186.
32. "Private Firms Take Over Public Functions: Germany, Switzerland," *Urban Innovation Abroad*, 4(9):1 (Sept. 1980).
33. Alan C. Miller, "Law Strips Guard Force at Airport of Authority," *Los Angeles Times*, 26 May 1987.

ers beyond that of citizen's arrest to be exercised in certain limited areas to situations where private officers have virtually full police powers and jurisdictionwide authority.

A Rand Corporation study for the Department of Justice defined deputization as "the formal method by which federal, state, and city governments grant to specific, named individuals the powers or status of public police [or peace officers]—usually for a limited time and in a limited geographic area."[34] As one student of the history of private police in the United States observed, "Certain classes of private police such as railroad detectives, campus security guards, and retail security guards may be granted special powers concerning arrest and search."[35]

According to a 1985 study for the Department of Justice, approximately 25 percent of medium-sized and large police departments deputize special officers or give them special police powers, probably more often to proprietary—that is, in-house—than contract—that is, involving an outside firm—security agencies.[36] Of proprietary security managers and contract security managers surveyed in 1984, 29 percent and 14 percent respectively, indicated that their personnel had special police powers.

New York City retail security officers illustrate a limited form of this type of privatization. The police department has authorized retail security officers in some establishments to "provide surveillance, make arrests, transport suspects to police holding facilities, complete records checks, and enter criminal history information."[37] Private security personnel exercising these powers must be trained in the apprehension of suspects and various legal issues. Private security officers in Washington, D.C., may also be awarded certain police powers of arrest. Those licensed to carry weapons are designated as "special police officers."[38] As of the early 1970s, the District had about 2500 such officers. In the 1970s, Boston passed an ordinance to establish training and clothing requirements as well as guidelines for the use of firearms for "special officers" who have the power of arrest on the employer's premises.[39] In Maryland, a law allows the governor to appoint "special policemen" with full police power on the premises of certain private businesses, and North Carolina has a similar law.[40] In Oregon, the governor can appoint "special policemen" in the railroad and steamboat industries. Texas permits its Department of Public Safety to commission "special rangers" who may work for private employers and "who have the full arrest and firearms powers of an official policeman and are empowered to enforce all laws protecting life and property."[41]

Another form of fully privatized police service occurs on the campuses of some private universities and colleges. Seven states have passed legislation giving some degree of police authority to campus security personnel at private

34. Sorrel Wildhorn and James Kakalik, *The Law and Private Police*, LEAA report R-872-DOJ (Washington, DC: Government Printing Office, 1971), 4:4.
35. Theodore M. Becker, "The Place of Private Police in Society," *Social Problems*, 21(3):446 (1974).
36. Cunningham and Taylor, *Private Security and Police in America*, pp. 40, 324.

37. Stewart, "Public Safety and Private Police," p. 761.
38. Athelia Knight, "'Rent-a-Cops' Pose Problems for District," *Washington Post*, 10 Nov. 1980.
39. Toni Schlesinger, "Rent-a-Cops, Inc.," *Student Lawyer*, Dec. 1978, p. 43.
40. O'Toole, *Private Sector*, p. 10.
41. Ibid.

universities.[42] In some states, private campus police may receive police powers via deputization by governors, courts, law enforcement agencies, or city governments. Campus police at the University of Southern California have been given certain powers of arrest under California Penal Code §830.7, pursuant to a memorandum of understanding with the Los Angeles Police Department. These powers grant the campus police at that university the same arrest powers as a California peace officer while they are on duty and within their jurisdiction and while responding to calls off-campus in the area surrounding the university.[43]

One of the purest forms of full police privatization is that of railroad police in a number of states. Enabling legislation was originally based on problems of interstate operation and the lack of public police protection in some areas. "In many parts of the country, the railroad police provided the only protective services until government and law enforcement agencies were established."[44] One Pennsylvania study, conducted in the 1930s, cited an 1865 state law that authorized railroad companies to employ security personnel who were also commissioned with full police powers in the county or state. If appointed by the state, railroad officers in Pennsylvania and New York, for example, maintained and exercised full police powers both on and off railroad property.[45] In fact, they were obligated to do so as sworn peace officers. Philadelphia at one time had hundreds of deputized, private security personnel providing police patrol services.

Fully private police services also exist on Paradise Island in the Bahamas. Virtually all island police activities are supplied by a firm employed by the island's hotels and resorts. The firm employs sixty to seventy guards and several vehicles, plus three or four administrative personnel and a captain. The company has responsibility for protecting 25 to 30 firms and the island territory. One analyst concluded that "since [the private security firm took] responsibility for protective services of the island, they have had one of the best records for low incidence of theft, rape, and assault in the area." This is in contrast to major incidences of theft and assault on the main island of New Providence, which has similar tourist attractions.[46]

Perhaps the foremost U.S. example of privately financed and provided police is that of the Patrol Special Police in San Francisco. Patrol specials are private individuals who undergo 440 hours of police academy instruction—the same as for reserve officers—and are sworn in as peace officers, one step below police officers. Once licensed, they are permitted to bid on one or more of some 65 private beats, with their salaries paid by merchants or residents along their beats. On the more lucrative beats, patrol specials hire assistants, who must complete the same training, and in some cases even hire security guards and administrative staff—thus, in effect, becoming mini police departments. They

42. Karen Hess and Henry M. Wrobleski, *Introduction to Private Security* (New York: West, 1982), p. 276.
43. Conversation with Lieutenant William Kennedy, University of Southern California, Campus Police, Jan. 1987.
44. Hess and Wrobleski, *Introduction to Private Security*, p. 13.
45. J. P. Shalloo, *Private Police* (Philadelphia: American Academy of Political and Social Science, 1933), pp. 25, 208.

46. James Gallagher, "The Case for Privatizing Protective Services," in *Facets of Liberty*, ed. L. K. Samuels (Santa Ana, CA: Freeland Press, 1985), pp. 93-96.

are legally members of the San Francisco Police Department, as reservists are, and are required to respond to police calls in their area.[47] Recently, however, the patrol specials have had to undertake a lawsuit to mandate California's Police Officer's Standards and Training Commission to continue their certification as sworn peace officers.

THE UNIQUENESS OF POLICING

Why does privatizing the police function seem to be so difficult? It cannot be for lack of traditional roots. In feudal Britain, the nobility hired others to discharge their required protective duties.[48] Some private law enforcement personnel were partially financed from fees for the recovery of stolen goods.[49] In London, some individuals eventually organized private patrols as a deterrent to crime and as a more formal means of pursuing criminals. During the 1700s and early 1800s, the British people were quite skeptical of a tax-supported government police for fear that it would provide political leaders with a means of oppression.[50]

The British system of requisitioning watchmen or using volunteers was transported to America. Gradually, social attitudes and norms accepted the institution of public police forces in major U.S. urban areas, as occurred in London.

47. Bill Wallace, "The Patrol Specialists—Salesmen with Badges," *San Francisco Chronicle*, 30 July 1984; idem, "Unique San Francisco Private Cops and How They Operate," ibid.; Christine Dorffi, "San Francisco's Hired Guns," *Reason*, Aug. 1979, pp. 26-29, 33.

48. Becker, "Place of Private Police in Society," p. 444.

49. Hess and Wrobleski, *Introduction to Private Security*, p. 9.

50. T. A. Critchley, *A History of Police in England and Wales* (London: Constable, 1967, 1978), pp. 29, 42.

With the advent of the first public police department in New York in the 1840s, other major cities quickly adopted public police departments. For over 100 years, most police protection in the United States has been publicly provided. Thus it is not surprising that many people now find it difficult to accept the notion of private police, despite their long history.

Another attitudinal factor is that some of the duties and activities undertaken by the police are qualitatively different from most other public services. The police have the right and obligation to use force, even deadly force, in the pursuit of their duties and receive special state authorization to do so.

A further barrier to police privatization is the difficulty of arranging for individuals to finance and consume police services, in contrast to other public services, privately. Moreover, even if a number of individuals in a neighborhood employ a private agency for patrol services, it is difficult to force those who do not wish voluntarily to finance these services to pay for the benefits they receive. One approach to dealing with this public goods problem is the deed-based, or mandatory-membership, home owners' association, in which security services are funded from mandatory membership dues. For example, in Los Angeles neighborhoods such as Bel Air and Beverlywood, a major fraction of the annual association dues goes to pay for the contract services of a private security company. This commercial firm provides 24-hour-a-day armed vehicular patrol by nonsworn security officers.

Another major barrier to police privatization is that of control and accountability. Understandably, people have serious concerns that those who are authorized to use deadly force be held accountable and under the control of the law. It certainly seems reasonable to require that all police personnel be properly trained and certified at the level of responsibility required by their particular duties. Moreover, as with any agency enforcing laws, strict quality-control regulation is appropriate, either through detailed contract provisions or regulatory oversight.

In addition to the public's natural apprehension, perhaps the greatest political barrier to privatizing police services is that of union opposition. As shown in the Reminderville and Oro Valley cases, public-police-officers' associations will strongly react to any local jurisdiction that attempts to privatize police services. But in light of today's budgetary constraints, it is shortsighted to permit special-interest political pressures to override the public's interest in cost-effective public services.

The final barrier to police privatization is legal restrictions. Upon close examination, however, these may not be as much of a problem as one would think. The attorney for Reminderville, Ohio, found that there was no state law preventing the contracting of police services. In the case of Troutman, North Carolina, the state deputy attorney general indicated that it was not technically illegal for the city to contract out its law enforcement, as long as the private police were sworn in as official police officers.[51] Perhaps, as in the Oro Valley case, it is more the political fight or potential legal costs that constitute the barrier.

CONCLUSION

On the one hand, there has been little progress in privatizing full or regular police services by U.S. local governments. To our knowledge, there is currently no city or county that is contracting for regular police services.

On the other hand, there does seem to be an increasing acceptance of more limited forms of privatization. The concept of special fees to beneficiaries in order to finance specialized police services, such as burglar alarm response, is increasingly accepted as fair and reasonable. Contracting out police support and ancillary services is growing steadily as well.

There are also signs of a gradual load shedding of certain police services to the private security industry, often through deputization or the award of special police powers. Assuming continued fiscal constraints on local governments and continued high levels of crime, it can be expected that police departments will gradually turn over more and more responsibility for law enforcement vis-à-vis private property to security organizations.

51. Conversation with Associate Attorney General Eddie Caldwell, Mar. 1984.

Book Department

PAGE

INTERNATIONAL RELATIONS AND POLITICS 119
AFRICA, ASIA, AND LATIN AMERICA 127
EUROPE 136
UNITED STATES 142
SOCIOLOGY.. 153
ECONOMICS 159

INTERNATIONAL RELATIONS AND POLITICS

COLLINGRIDGE, DAVID and COLIN REEVE. *Science Speaks to Power: The Role of Experts in Policymaking.* Pp. xi, 175. New York: St. Martin's Press, 1986. $27.50.

HISKES, ANNE L. and RICHARD P. HISKES. *Science, Technology, and Policy Decisions.* Pp. ix, 198. Boulder, CO: Westview Press, 1986. $35.00. Paperbound, $15.95.

Our society clings to the misperception that science deals exclusively in the realm of facts and rational thinking while politics consists of opinions, emotional arguments, and positions arrived at by a careful calculation of self-interest. Science seems to be everything politics is not. According to this reasoning, public policy could be considerably improved by generously substituting the pristine scientific approach for the messy approach of politics.

Of course, it does not work that simply. The "double wedding of science and government, knowledge and power," in the words of Hiskes and Hiskes, has resulted in what Collingridge and Reeve call an "unhappy marriage." Both books do a superb job of explaining why. Their perspectives differ. Collingridge and Reeve look at how science influences policy. Hiskes and Hiskes turn it around and look at how policy shapes the advancement of science and the application of technology.

Their audiences also differ. Hiskes and Hiskes have written a textbook consisting of a general theoretical model and several cases, such as nuclear power, energy, and biomedical technology, each of which is analyzed using the model. Collingridge and Reeve address an audience perhaps more familiar with the tension between science and government. Theirs is a more elegant argument—more theoretical and sophisticated.

Both discussions, however, can easily go beyond their obvious markets. Hiskes and Hiskes do not gloss over the complexity of their subject. Collingridge and Reeve avoid the trap of specialized jargon; the general contours of their argument can be easily understood and appreciated.

Despite its brevity, *Science Speaks to Power* is a rich and insightful discussion. The many contributions cannot be conveniently summarized, but two points should convey the tenor of the authors' approach. First, they convincingly demonstrate that involvement in policy debates compromises the conditions for pure science. Those conditions—an autonomous and specialized scientific community, and a "low level of criticism" toward consensual theories of science—are virtually impossible to achieve when science mingles with policy.

Second, they debunk the myth that science will somehow bring political debates closer to consensus. On the contrary, policy questions exacerbate differences among experts. When science becomes associated with policy, the consequences of scientific error become greater. An academic disagreement now has real implications for the distribution of wealth and power. The search for error intensifies. Instead of promoting consensus, scientific experts can easily make the debates more adversarial.

Collingridge and Reeve offer the best articulation of a contentious theory about the relationship between science and policy. Some readers will consider it exaggerated; I prefer to consider it a straightforward presentation of an important and plausible line of reasoning.

Hiskes and Hiskes share the skepticism of Collingridge and Reeve that political discourse can somehow be made simpler by science and technology, or that policy disputes arising from technological applications are best left to the engineers who have created the technology. Policy disputes over technology are unavoidably political; they are not technical. Further study of the consequences of acid rain will not change the fact that this is a question of distributive justice, requiring political decisions about who will bear the cost. A greater technical understanding of genetic engineering will not reduce the need to make difficult ethical decisions about what is meant by life.

Hiskes and Hiskes contend that decisions about science and technology must be made democratically. In each of the cases reviewed in their book, they show how the public can be involved in the formation of policy. This is often more problematic than they acknowledge, but like Collingridge and Reeve, they are developing an important thesis, one especially appropriate for students to confront.

Scholars of science and government have been rewarded with two excellent books. The topic will grow in importance as scientific advances raise increasingly serious policy questions. Hiskes and Hiskes offer students an intriguing introduction to the complexity of this policy arena. Collingridge and Reeve construct an argument that should engage scholars concerned not only with science and technology but with decision making in general. The strength of the two books is not that they present elaborate and original research; rather, they offer a perspective and an approach to help us understand the character of the relationship between science and government.

KENNETH P. RUSCIO
Washington and Lee University
Lexington
Virginia

DORRIEN, GARY J. *The Democratic Socialist Vision*. Pp. xi, 180. Totowa, NJ: Rowman & Littlefield, 1986. $24.95.

Gary J. Dorrien analyzes the democratic socialist movement in the First World, where Marxist revolutions were theoretically supposed to occur, with an emphasis on its religious roots. The reader might feel forewarned that he or she is in for another airy essay, but Dorrien offers "a chastened socialism without illusions."

His is the socialism of the Frankfurt Declaration of 1951, enhanced by such recent additions as the Swedish Meidner plan, which uses worker funds to transfer ownership of corporations to workers and provides a system of wage restraints to keep Sweden competitive in exports. These possibilities are open to American unions, Dorrien argues, whose pension funds constitute some 25 percent of the stock on the American stock exchanges. This is one of several well-reasoned arguments that democratic socialism is a viable option for the United States.

Dorrien focuses on William Temple, archbishop of Canterbury and a harbinger of the World Council of Churches, and on two modern secular socialists who owe a debt to religious principles, Norman Thomas and Michael Harrington. An ethical democratic socialism is opposed to a nonethical Marxism-Leninism, although it is perhaps too sweeping, given the entire corpus, to say that "Marxism . . . repudiates all forms of moral discourse." Temple is not so well known to many socialists, but Dorrien also manages to cast light on the intellectual development of the better-known Thomas and Harrington. The coverage of Harrington includes a challenging discussion of Marxism and a fine coverage of the fierce debate on the Left in the 1960s concerning communism and U.S. intervention in the Third World. Surely Dorrien is among a minority who are aware of—or admit—the fate of the Vietnamese Trotskyists.

Dorrien also takes on Michael Novak, a substantial critic of socialism whose intellectual roots are close to the religious Left. Novak proposes democratic capitalism, to which Dorrien poses two telling criticisms: it is imperialistic, and it is not democratic enough; in particular, Novak ignores economic freedoms.

A perennial weakness of democratic socialism as proposed by Dorrien is the tacit assumption that it implies support of a broad agenda of issues that are individually challengeable in their own right. It is not obvious, for instance, that social democrats should automatically oppose all U.S. interventions in the Third World.

Dorrien's very readable book is highly recommended to the lay reader interested in the ideas behind political controversy and to students of political philosophy. Another intended readership is those Americans for whom "socialism" is still an ominous and misunderstood term.

WILLARD D. KEIM
University of Hawaii
Hilo

FREIDMANN, JOHN. *Planning in the Public Domain: From Knowledge to Action.* Pp. xii, 501. Princeton, NJ: Princeton University Press, 1987. $55.00. Paperbound, $16.95.

Many social scientists are critical of traditional planning theory and practice as developed in the United States. Marxists argue that mainstream planning is a repressive instrument of the bourgeois state that merely seeks to ameliorate the contradictions of capitalism. Others writing from a hermeneutic perspective contends that state-orchestrated planning is excessively structural, ignoring the particularistic meanings that individuals attach to social situations. In this path-breaking book, John Friedmann draws upon Marxist and hermeneutic approaches to offer an original and compelling new model of radical planning.

Friedmann begins his analysis with a brief but helpful overview of planning concepts and theory. He maintains that the purpose of planning is to link knowledge to action in the context of a public domain. Knowledge can be linked to action through either societal guidance, where planning is practiced by an elite in support of the state, or social transformation, where planning is democratized and practiced by ordinary citizens apart from the state. Furthermore, planning theory varies according to political ideology. Conservative theory accepts the current capitalist mode of production while radical theory is oppositional, seeking to restructure capitalism.

Friedmann next offers a lengthy historical review and critique of four current models of planning theory: (1) policy analysis, which stresses societal guidance and conservative political ideology; (2) social reform, which stresses societal guidance and radical political ideology; (3) social learning, which stresses social transformation and conservative political ideology; and (4) social mobilization, which stresses social transformation and radical political ideology. Policy analysis developed from the earlier work of systems engineers, such as Wiener; neoclassical economists, for example, Hayek and Arrow; and public administration theorists, for example, H. Simon and Lasswell. Social reform is an outgrowth of the writings of sociologists, such as Saint-Simon, Comte, Weber, and Mannheim; institutional economists, including Tugwell and Galbraith; and pragmatists, for example, James and Dewey. Social learning emerged from scientific management, such as Taylorism, and organizational theory, as in the work of Mayo, Bernard, and, most important, Lewin. Finally, social mobilization emerged from the work of historical materialists, such as Marx and Engels; utopians, like R. Owen and Fourier; and anarchists, such as Proudhon, Bakunin, Kropotkin, and Sorel. Friedmann masterfully discusses most of these theorists and perspectives in his survey of the history of planning theory.

In the last part of the book, Friedmann presents his own model of radical planning. He contends that traditional planning approaches, especially policy analysis, cannot address the global problems precipitated by rapid social change and an irresponsible and oppressive capitalist world system. Because they are allied with the state, traditional planners cannot fundamentally critique and transform the state. The result is a crisis in planning, where the ineffective link between knowledge and action allows world problems to worsen. Friedmann's solution to this crisis is to recenter planning apart from the state in civil society, so that planning occurs by "ordinary citizens" in support of a "public domain" of common interests and a common good. "Planning in the public domain"—or radical planning—is organized, on a micro level, around households and regional action groups and, on a macro level, around peasant societies and the global community. Radical planning effectively mediates knowledge and action, becoming an instrument that empowers citizens in the struggle for social transformation.

While Friedmann's analysis is brilliant and provocative, there are still a few shortcomings with the book. For example, occasionally Friedmann tacitly acknowledges that pragmatism might be considered either a social reform or a social learning theory of planning. The difficulty in locating pragmatism illustrates the somewhat simplistic nature of Friedmann's classification scheme. Furthermore, there are some curious omissions in the book—scant attention is given to B. F. Skinner and behaviorist psychology, Paulo Freire and critical pedagogy, and Ebenezer Howard and the garden cities movement. Each of these theorists and perspectives deserves discussion in a comprehensive historical survey of planning. Finally, some may criticize Friedmann's notion of planning in the public domain as merely old mainstream planning wine in a new radical planning bottle. In particular, Marxists consider talk of common interests and a common good within a capitalist world system to be illusory.

Despite these criticisms, *Planning in the Public Domain* is almost certain to become a social science classic. I recommend this book to any reader seriously concerned about posterity.

RICHARD A. WRIGHT
Kansas State University
Manhattan

HUTCHISON, WILLIAM R. *Errand to the World: American Protestant Thought and Foreign Missions.* Pp. xii, 227. Chicago: University of Chicago Press, 1987. $24.95.

With the appearance of ground-breaking books by John Fairbank, Valentin Rabe, Joan Brumberg, Jane Hunter, and others, and now with the publication of this more synoptic book by William R. Hutchison, the history of the American Protestant missionary movement has come of age, superseding the hagiographical and debunking phases of missionary history with a recognition of the cultural and political importance of this movement both for the United States and for the world. For long periods of our national history, much of what Americans heard about a wider world, especially the more remote lands of Asia, Africa, and Oceania, was from the dramatic reports of missionaries. Enthusiasm for foreign missions among a large segment of the American Protestant public, and the experience of sending missionaries and receiving them back, were important influences in the forming of American internationalism. And the educational and cultural impact of the missionaries abroad was far from negligible.

Hutchison has capped many earlier studies by others with an analysis of what American missionaries and missionary theorists thought about what they were doing in their missionary enterprises and how these thoughts intertwined with important themes in the cultural self-understanding of Americans. Thereby he has uncovered a neglected element in American intellectual history.

In a series of chapters that move from Roger Williams and John Eliot to the latest missionary theorizing among ecumenical church leaders, Hutchison explores a major theme: to what extent can missionary labors be separated from the American cultural baggage of the missionaries? Far from this being only a recent concern, he shows it to have been a pressing issue from the beginnings of the movement, making us aware of the sophistication with which the subject was handled by such pioneer missiologists as Rufus Anderson. Other relatively unknown but important figures who thought about this issue are also brought to our attention, such as Daniel Johnson Fleming, but one also sees such familiar names as those of the great missionary statesmen—Mott, Speer, Eddy—placed into context. As the title suggests, Hutchison places the missionary movement in the context of the image of an errand, a reversal of the Puritan errand to America, as Americans engaged in the errand of taking America's Christian Zion abroad. Aware of the profound influences of American culture upon the movement, and its inevitable relationship to American imperialism in the widest cultural sense of that word, Hutchison nevertheless avoids a reductionism that would ignore genuinely religious motives and interpret the movement merely as imperialist ideology.

Errand to the World has many other strengths. Fundamentalists and recent conservative Evangelicals are taken seriously as participants in the discussions of mission theory; the final chapter provides a world context for an American discussion; and a reflective brief afterword suggests Hutchison's serious engagement with his subject. It is an important book on an important subject.

DEWEY D. WALLACE, Jr.
George Washington University
Washington, D.C.

LAIRD, ROBBIN F. *The Soviet Union, the West, and the Nuclear Arms Race.* Pp. xii, 236. New York: Columbia University Press, 1986. $35.00.

LEBOW, RICHARD NED. *Nuclear Crisis Management: A Dangerous Illusion.* Pp. 226. Ithaca, NY: Cornell University Press, 1987. $24.95.

Robbin Laird's extremely well-organized study provides an outstanding survey and evaluation of the changing nuclear weapons policies of the two superpowers, the Soviet Union and the United States, and of the West European nuclear powers, France and Britain. A well-known defense analyst and expert on Soviet and West European affairs, Laird discusses in the first part of the book the development of and changes in Soviet nuclear policy and its usefulness for Soviet policies designed to cause and increase tensions within the Western alliance. The second section presents an analysis of American nuclear policy, the most current U.S. strategic defense options, and the prediction that the U.S. defense will in the future depend on mobile intercontinental ballistic missiles or active defense of missile silos and command posts.

France's nuclear modernization program is markedly influenced by Paris's perceived need to expand the concept of independence and of maintaining the myth that French territory is "sanctualized."

In 1980, the Conservative government of Britain decided to replace its aging Polaris force with the U.S. Trident submarine, and in 1982, it was decided to arm the submarines with D-5 missiles. The strong rejection by the parliamentary opposition and various peace groups caused the implementation of the program to be questioned at times. Since the publication of this book, the Thatcher government has received a new mandate by the electorate and, therefore, there seems to be less doubt about the introduction of the Trident.

The final, and probably most challenging, part of the book examines Soviet political military strategy, which attempts to take into account the existing strategic nuclear parity, the Euromissile competition, and France's "independent" security policy. Moscow's "anti-coalition strategy" toward the Western alliance is directed at diminishing the cohesion of the North Atlantic Treaty Organization (NATO). The Soviet Union, according to Laird, believes the post-World War II "Atlanticism," that is, America's strong influence over political and economic developments within Western Europe, has become less significant. As a matter of fact, Moscow sees in the emergence of West European and Japanese economic power the development of "three centers of imperialism," marked by increasing "interimperialist contradictions."

The nuclear arms issue and its divisive influence on the Western alliance—for example, the present concerns of West European governments about the elimination of the intermediate-range nuclear forces—are by far not the only means used by Moscow to destabilize the alliance and Western societies. Soviet overall strategy is to change the balance of forces—in Soviet terminology, "the correlation of forces"—in favor of the world socialism.

Richard Ned Lebow's book reflects his concern about the capability of the superpowers to manage a major confrontation and thus avoid the escalation of the crisis into a nuclear war. His previously held chair at the National War College and especially his position as scholar in residence in the Central Intelligence Agency provide him with excellent opportunities to observe and evaluate the American command and control apparatus; as far as the Soviet Union's situation in this respect is concerned, Lebow depends on Soviet published material. It would be most difficult for someone with less intimate exposure to the inner workings of the existing American organizational and jurisdictional structure, designed to operate during major crisis situations between the superpowers, to question his rather pessimistic evaluation of American capabilities to keep a crisis under the nuclear threshold.

Lebow seriously questions if the present command and control systems of both powers would succeed in preventing a nuclear confrontation as they were able to do in the 1962 Cuban missile crisis.

Lebow draws several conclusions from our detailed knowledge of the various events, policies, and contemporaneous military strategic dogmas that led to World War I. He believes that the same categories, preemption, loss of control, and miscalculated escalation, are equally significant in contemporary crisis situations.

In his final chapter, Lebow provides a number of recommendations intended to enhance crisis stability of the command and control structures, assuring them at least "short-term" survival in case of an exchange of nuclear missiles, a vastly better-informed political leadership, and key personnel better prepared to perform under stress. Emphasis is placed on the enormous psychological stress to which the leaders would be exposed in a nuclear crisis because of their awareness of the unprecedented destructiveness of nuclear weapons.

I have difficulties in agreeing with Lebow's statement that "war-threatening crisis may actually have increased as a result of the demise of détente and the revival of the cold war." It depends on what is understood to constitute détente. This period in post-World War II developments, as seen by many observers, was also the era during which the Soviet Union carried out a massive arms buildup while the Western democracies indulged in wishful thinking.

Lebow credits the Soviet leadership with the same motivations and considerations as those ascribed to the elected leaders of Western democracies. The lessons to be learned from the pre-World II era are frequently overlooked, that is, that a leadership—in this case, Hitler—determined to pursue an expansionist policy is not motivated by defensive security considerations. An analysis of the events leading to World War II is far more relevant to the present situation than the July crisis of 1914.

Lebow's statement that "although the United States is publicly committed to ride out a first strike, it is *not* committed to refrain from the first use of nuclear weapons" requires some clarification. He obviously distinguishes between an all-out nuclear first strike by the United States against Soviet territory and the first use of nuclear weapons by NATO forces in response to a full-scale assault by Warsaw Pact conventional forces in the European theater. In the NATO case, the threat of the first use of nuclear forces had served so far as an effective deterrent. This concept of "extended deterrence" was also the rationale for American nuclear forces during the 15 years when the Soviet Union could not mount a large-scale nuclear attack on the United States.

Lebow strongly favors arms control as a significant stabilizing factor and consequently supports the Anti-Ballistic Missile Treaty. In spite of Soviet superiority in intercontinental ballistic missiles and Soviet strategic defense installations, he opposes ballistic missile defense, the aim of President Reagan's Strategic Defense Initiative, as well as the MX missiles because they "would intensify mutual pressures to preempt in a crisis."

ERIC WALDMAN
University of Calgary
Alberta
Canada

LUTTWAK, EDWARD N. *Strategy: The Logic of War and Peace.* Pp. xii, 283. Cambridge, MA: Harvard University Press, 1987. $20.00.

LEVITE, ARIEL. *Intelligence and Strategic Surprises.* Pp. xiii, 220. New York: Columbia University Press, 1987. $27.50.

Over the last eight years, the military buildup in the United States has been subject to constant criticism from those who believe that the acquisition of large numbers of new weapons has been accomplished in an atmosphere either devoid of strategy or where strategy has been consciously altered to require continued defense spending. Edward N. Luttwak does not directly enter this debate with his new book, *Strategy: The Logic of War and Peace,* but he does such a masterful job of putting the concept of strategy in an understandable context that this book cannot help but affect the future course of U.S. defense.

Luttwak says, "My purpose has been to uncover the workings of the paradoxical logic . . . offering in the process a general theory of strategy that describes but does not prescribe." Greatly simplified, the general theme of this important book is that actions must be considered in light of the likely response they will elicit from an opponent. And while this sounds obvious, Luttwak demonstrates that, for every decision, each of the five levels and two dimensions of strategy contains at least one complicating paradox.

It is unfortunate that our military and political leaders are not schooled in strategy as Luttwak defines it. In fact, by his definition, it may be that good strategists are born, not made. As it now stands, this book has been written for a decision maker who does not exist, and it will be read by decision makers who are, more often than not, chosen for the very characteristics that prohibit the type of thought advocated by Luttwak.

Luttwak recognizes this point, and he concludes that "it would be hard for democratic political leaders to follow policies that can so easily be branded as illogical and contradictory." For this reason, an overall, grand strategy for a nation may not be good because it could stifle the paradoxical approach that denotes good strategy—an interesting concept given the criticism of the Reagan administration on this very point.

This book is particularly good in its discussions of armed suasion and the development, on the part of the opponent, of a mindset that creates deterrence. Luttwak does not end with a prescriptive method for creating strategy. Instead, he demonstrates why any fixed methodology must fail, and he tries, instead, to create an awareness of the process that leads to successful strategy.

This is an excellent, thought-provoking book. Luttwak's writing style is difficult in the theoretical passages, but the examples are clearly explained and they provide exceptional illustrations of the point being made. The notes that accompany each chapter are a major addition to the book and are needed to put the examples in context—it is unfortunate that they are gathered at the end of the book where they are so inaccessible. In sum, this is a book well worth the almost ten years it has taken to complete it. It is required reading for anyone interested in the defense of the United States.

Intelligence and Strategic Surprises by Ariel Levite has the primary purpose of trying "to improve the understanding of strategic surprises and related phenomena such as threat perception and response to threat." In this respect, it only partially fulfills its goal. The book is written in a clumsy, dissertation style with innumerable paragraphs where Levite explains what he is going to tell us or what he has just told us.

These untidy organizational characteristics are further complicated by the tone of the book, for example: "I wish to point out that my . . . novel conclusion . . . "; or "my in-depth study . . . permit[s] me to . . . offer an . . . explanation . . . that is better than alternative explanations"; or "I now . . . review all the sources of intelligence on Japan possessed by U.S. intelligence prior to that day." All these things, when combined with Levite's tendency toward repetition, make reading this book a tedious and aggravating assignment.

The first 38 pages of the book are devoted to a review of the literature of the field. This is followed by a 56-page, extremely detailed review of the events leading up to the Pearl Harbor attack. Unfortunately, Levite's real agenda throughout these two sections seems to be a frontal assault on the work of Roberta Wohlstetter, whose 1962 *Pearl Harbor: Warning and Decision* remains the classic work in this area. Whatever the merits of Levite's arguments, they belong in either a journal article or a specialized book devoted to this debate, not as the unstated side issue in a general book on strategic surprise.

Levite then discusses the battle of Midway at some length. He uses this battle as a companion case to the Pearl Harbor study to show a situation where intelligence succeeded in avoiding a strategic surprise. While he goes to great lengths to defend the legitimacy of this choice, I am still troubled by the use of two cases that span the structural changes brought on by the declaration of a major war.

The last two sections of the book discuss the theory of warning, threat perception, and response, and these discussions are illustrated with many current examples. In spite of their obvious link to the topics covered in the first part of the book, these sections form such a clean break with the chapters that preceded them that they almost stand alone. Neither of these chapters is particularly innovative, but they are interesting and are by far the best part of the book.

In sum, this book will be of interest only to a very small group of scholars who study the theoretical aspects of intelligence gathering and dissemination. For these individuals, the book holds a great deal of solid research and information. It also provides a number of conclusions and statements by Levite that will undoubtedly prove to be the subject of many debates in the future.

WILLIAM WEIDA

Colorado Springs

SCHALL, JAMES V. *Reason, Revelation, and the Foundations of Political Philosophy.* Pp. 254. Baton Rouge: Louisiana State University Press, 1987. $27.50.

The centuries-old scholarly dialogue with the Greco-Roman and medieval philosophers is alive and well, as evidenced by this latest work of James V. Schall, assistant professor of government at Georgetown University. But what Condorcet and other modernists saw as the war between philosophy and superstition Schall sees as the war between shallow ideologies and true philosophy.

The book's argument unfolds in eight chapters and a brief introduction and conclusion. The weakest of the chapters is the first, "The Statement of the Problem of Political Philosophy," while the strongest is the last, "Jerusalem, Athens, Rome." Three main purposes seem to structure the analysis: to prove that modern political philosophy has usurped the role of Christian metaphysics; to argue that the ultimate solutions to the deepest human problems lie beyond the political sphere; and to demonstrate that political philosophy stands today in dire need of a return to the revelational traditions of Judeo-Christianity, along with a reintegration of classical political thought.

Schall's book has a number of solid virtues that commend it to the thoughtful reader. It is written in a fluent style with minimal obfuscatory jargon. It is tightly and cohesively organized. Its basic themes are forcefully and consistently argued throughout. And it is heuristic in the sense of stimulating thought, provoking discussion, and questioning our age's *Zeitgeist*.

Yet despite these strengths and others, I found the book deficient in a number of key areas. It goes little if any beyond the work of Gilson, Strauss, Arendt, and Voegelin. It leaves imprecisely defined such crucial terms as "ideology," "political metaphysics," and "gnosticism." It offers a highly one-sided interpretation of classical philosophy as it relates to Christian revelation. It denigrates modern politics and philosophy without providing a corresponding analysis of the faults of their revelational counterparts; are there none such? It is, in fact, more a study in the philosophy of history than a work of political philosophy per se. Finally, though the book makes it clear that revelational political philosophy would raise different questions from secular philosophy, it is not obvious that the results would be all that different in policy terms. In short, Schall fails to show that politics and public policy would be better in any real sense if they became less anthropocentric and more theocentric, cosmocentric, or Christocentric.

Still, every card-carrying, philosophically inclined scholar should read this book, as much for its belletristic grace as for its challenging themes. Moreover, let it be noted here that the book's felicitous writing and its crisp, intelligent commentary exhibit the enduring advantages of a Catholic education in a putatively decadent age.

FRANCIS M. WILHOIT
Drake University
Des Moines
Iowa

AFRICA, ASIA, AND LATIN AMERICA

AMATE, C.O.C. *Inside the OAU: Pan-Africanism in Practice*. Pp. xiv, 603. New York: St. Martin's Press, 1986. No price.

Although the Organization of African Unity (OAU) is approaching its twenty-fifth anniversary, until now there has not been a comprehensive history of the institution, nor has there been a probing attempt to analyze the dynamics of the organization. *Inside the OAU* is a herculean attempt to fill this gap in knowledge.

The book is a well-written, comprehensive history intended to give readers an insider's view of the origins and evolution of the OAU. The volume is divided into three major sections. Part 1 deals with the historical antecedents of the organization, tracing its roots from the first Pan-African Congress in 1900 to the actual declaration of the OAU in 1963. A very lucid account is presented of the dynamics of the processes and personalities involved in the creation of the organization. Subsequent chapters in part 1 describe in detail the functional units of the OAU secretariat.

The remainder of the book focuses on the internal workings of specialized as well as ad hoc committees. Rich description and numerous examples are used in an effort to give a comprehensive picture of the day-to-day workings of the OAU. Amate paints a sympathetic picture of an institution faced with the monumental task of serving as an agent of unity, conciliation, and cooperation for the 51 member states and of presenting the posture of a united Africa when dealing with the rest of the world. All of this must be accomplished with limited monetary and technical resources. Amate repeatedly makes the point that it is remarkable how efficiently the OAU operates in spite of its resource constraints.

Amate sets out to paint the big picture of the OAU, and to a large extent he succeeds. The strength of his work lies in its breadth of coverage and its comprehensiveness. Amate also seeks to present a balanced assessment of the organization; however, he is less than successful in meeting this goal. The book is filled with rich description but lacks critical, penetrating analysis. Amate's bias is all too evident, detracting from the claim to balance.

A second strength of the book is the thorough and systematic research that was involved in crafting this study. Amate combed and sifted through an enormous amount of primary data on the OAU and also conducted in-depth interviews with key bureaucrats inside the organization. The data are well organized and skillfully presented albeit often in a rather wordy fashion. If Amate's research is a strength, however, his method of documentation is grossly lacking. The volume has not a single footnote. Consequently, it would be difficult for the reader to follow up and check some of the empirical assertions made in the book. Moreover, the power of the book as a reliable encyclopedic reference is significantly lessened.

Despite its shortcomings, there is no book like *Inside the OAU*. It is must reading for anyone who seeks to understand the OAU.

EDMOND J. KELLER

University of California
Santa Barbara

BARNES, SANDRA T. *Patrons and Power: Creating a Political Community in Metropolitan Lagos.* Pp. x, 261. Bloomington: Indiana University Press, 1986. $32.50.

Sandra T. Barnes's compelling study of patron-client relations in Mushin, a growing urban community in metropolitan Lagos, is not only an exemplary case study of the creation of a political community but also a significant contribution to the analysis of political patterns in contemporary Africa. *Patrons and Power* investigates the ways in which nonelite residents, with little political access, devise methods of acquiring political status, enhancing their mobility, and securing their position within the public arena. The linkage between informal political processes and formal political institutions therefore stands at the center of this undertaking. Clientelism is conceptualized in these pages as the main method of tying together various spheres of political activity and consequently of connecting leaders and followers, rulers and ruled. This book, then, is concerned with power in its most basic sense: with how it is gained, how it is used, and how it is maintained.

Relying on materials gathered during extensive fieldwork in Nigeria, including a rich array of in-depth interviews, surveys, questionnaires, and archival documentation, Barnes commences her analysis with a historical account of the development of Mushin, one that highlights the diversity of its growing population and the official neglect it experienced throughout the colonial period. Administrative procedures were developed by a new crop of leaders that emerged within the changing urban context.

The making of these authority figures is discussed in two separate chapters, the first of which deals with the material base of political aspirants. Land ownership is portrayed as a key means of obtaining status within the community; building a house on these plots is seen as a crucial way of amassing wealth. Barnes's insistence on the centrality of the acquisition of autonomous resources in the process of leadership formation challenges the growing tendency to associate power in Africa with state-linked political office and suggests an important distinction between the material bases of formal and informal authority on the continent.

Chapter 4 deals with the second prerequisite of leadership: political skill. The politics of residence demanded the elaboration of normative codes of reciprocity, mutual obligation, and trust. The abilities of political aspirants to use their resources, connections, knowledge, and organizing capacity to mediate disputes, dispense benefits, and act as middlemen constituted latent but vital mechanisms for power accumulation. The inequality of patron-client ties was therefore mitigated by their reciprocal underpinnings—the position and power of leaders tempered by their need for legitimacy. The delicate balance between material and moral preconditions for the exercise of power help to explain both the range of leadership possibilities and the constraints placed on leaders' power. This analysis constitutes an important corrective to the tendency to view contemporary political leadership in Africa in purely personal and idiosyncratic terms, overlooking the rules that guide political transactions.

The analysis of the historical patterns of leadership formation lays the foundation for the in-depth examination of how power relations are organized. Barnes devotes separate chapters to chieftaincy, formal institutions, and factions as vehicles for the consolidation of leadership. In her discussion, she highlights the ongoing significance of chiefs as sources of authority and focuses of representation. She also stresses their changing position in the political realm: their durability is attributed to their flexibility in merging communal roles at the local level with interest group activities in the broader political system.

The final stage in the acquisition of power lies in its institutionalization. Bureaucratic approval for chieftaincy titles and for representation in governing bodies was the key to power perpetuation. Barnes astutely demonstrates how local patrons, enjoying grassroots support, vied for official recognition to cement their positions. In this process, they forfeited some of their local autonomy in order to secure their place in the wider political hierarchy. At the same time, a more cohesive power apparatus was put in place, one that sustains itself despite constant changes in the composition of those holding office at the top.

In the final chapter, Barnes analyzes the relationship between power and clientelism. She shows that patron-client ties are significant mechanisms for the expansion of social opportunities, the organization of diversity, the attenuation of ethnic and class cleavages, and the maintenance of order. Her analysis underlines the connection between hierarchy on the one hand and reciprocity on the other. The centralization of power is thus integrally associated with the creation and entrenchment of power networks. Barnes rightly places power in these aggregations.

Patrons and Power is a well-crafted, lucidly written study that challenges several fundamental tenets of contemporary African political studies. First, it emphasizes the creation of middle-level authority figures where many observers thought that a power vacuum had existed. Second, it questions the disjuncture between local and national politics that permeates the literature, pointing consistently and convincingly to the multiplicity of linkage networks that exists on the African political landscape. Third, through the accentuation of the norms of accountability that underlie patron-client links, it dispels the tendency to associate patronage with exploitation and clientelism with departicipation. Fourth, it documents the multiple possibilities for political representation and action in ostensibly closed political systems. Finally, it furnishes a vital antidote to the propensity to view African political power in hegemonic, indivisible terms, thereby expanding the field of political vision beyond the formal and the visible.

Barnes's superb study therefore raises a host of questions that require further attention. While Barnes has shown how macro political processes impinge on local politics and shape the choices available to aspiring politicians, she has not addressed the issue of how these local processes consequently affect national politics. If, indeed, local authority figures are co-opted into the formal system and a modicum of power is thereby extended beyond the political center, what are the ramifications of these trends for the structure of power on a countrywide basis? Is power in fact undergoing a process of reaggregation? What is its significance for shifts in regime forms and patterns of rule? In the same vein, does the institutionalization of patrons provide the basis for the consolidation of a new set of informal leaders? What is their relationship to established clientelistic networks? Finally, can alternative, nonpersonal methods of political action be established in such circumstances? By what means and with what results?

It is the mark of an exceptional academic work that it raises a host of new questions, thereby setting the parameters of an innovative research agenda. Sandra Barnes's analysis of patrons and power in Mushin, in its careful attention to detail and process, is a rare example of such a study.

NAOMI CHAZAN

Hebrew University of Jerusalem
Israel

BURKI, SHAHID JAVED. *Pakistan: A Nation in the Making*. Pp. xiii, 226. Boulder, CO: Westview Press, 1986. $28.00.

Political scientists who teach courses on South Asia have long awaited an integrated, single-author survey of Pakistan to match the widely used textbook by Hardgrave on India. This short volume by Burki partially meets this need. One of Pakistan's leading social scientists, employed since 1974 by the World Bank as an economist, he gives us an insightful and accurate but historical narrative of the country's origins and politics from a Pakistani point of view. The chapters display great interpretive skill rather than the analytical and comparative categories of contemporary political science. For instance, he points out the greater development of east-west rather than north-south communications in the Indus Valley, seldom mentioned in accounts of Pakistan. By calling Pakistan unique, however, he, like a historian, mostly misses potential fruitful comparisons with Israel, the other religiously based state in the world, in contrast to "secular" India.

Burki's analysis of his country's relatively high level of economic development—"a middle-income country"—and low level of social development—"very backward in health and education"—in chapters 3 and 4 reflects his professional expertise better. What is missing is a chapter devoted solely to the very threatening problem of national integration. Burki acknowledges the fear of Punjabi domination by the smaller provinces, but the very subtitle of the book, "A Nation in the Making," demonstrates a certain Punjabi insensitivity to the basic question of whether Pakistan is indeed a nation or, if not yet, is headed in that direction. Sometimes the reader would not know that he or she is learning about the same country as the one described by critics like Alavi, Ayoob, and Bin Sayeed. Thus he stoutly asserts that the benefits of the green revolution and emigrant remittances from the Arabian Gulf have been widely distributed through the classes.

Regarding Pakistan's foreign relations, Burki provides in chapter 5 a useful corrective to the pro-Indian bias of most American specialists and journalists on South Asia. He properly underlines the parallel between the princely states of Hyderabad and Kashmir, both of which India obtained by force. But Burki treats the Simla agreement negotiated by Bhutto with India in 1972 as tantamount to a "quasi-permanent" solution of the Kashmir dispute in India's favor as well as an acceptance by Pakistan of India as the dominant power in South Asia, conclusions that have been thrown in doubt again by the renewal of American military aid to Pakistan in the 1980s.

Burki says that his main thesis is that the purpose for which Pakistan was founded, a haven for Muslims from Hindu domination, is already achieved and that new goals need to be found. In conclusion, he finds these, surprisingly for an economist, in the definition of the country's political, social, and economic objectives in Islamic terms.

     THEODORE P. WRIGHT, Jr.
State University of New York
Albany

DAVIS, LEONARD. *The Philippines: People, Poverty & Politics.* Pp. xx, 225. New York: St. Martin's Press, 1987. $27.50.

This book, written by an academic currently with the Department of Social Administration at the City Polytechnic of Hong Kong, was not written for an academic audience. It is not footnoted and the two-page bibliography makes no reference to the works, major or minor, on Philippine people, poverty, or politics. Davis explains in the preface that he was working primarily from "unmarked documents and unreferenced photocopied articles, circulated in haste to groups of church workers and others."

What we do have in this book is a sympathetic, secondhand account by a foreigner of the repression, brutality, and poverty that the poorer Filipinos, both rural and urban, face on a daily basis.

The first two chapters set the context, with a quick discussion of the Philippines' history, economy, and relations with the United States. There is also a brief introduction to the customs and values of the Filipinos.

The heart of the book is the next six chapters entitled, respectively, "Poverty, Sickness and Disease"; "Prostitution"; "Persecution and Oppression"; "1985: The Child, the Priest, and the Crowd"; "Resistance and Revolution"; and "Certainty and Uncertainty." These chapters provide accounts of daily life told to Davis directly or as related to him by church workers on behalf of peasants, workers, fishers, and their families. What emerges from these stories is the wrenching human cost of the U.S. military presence in the Philippines, the militarization of the Philippines by its own leadership, and the continuing economic and political crises of the nation. Chapter 8 and a postscript give a quick overview of events up through about the middle of 1986.

Despite being someone who writes with a great deal of empathy for the plight of the poor, Davis does not appear to have learned to speak any of the Philippine languages. Neither he nor the editors caught a number of errors in the spelling of Filipino names, in the translations of Pilipino terms to English, and in the spelling of words in Pilipino.

The book is strongest at providing the reader with a window on the lives of average Filipinos as well as their persistent efforts to make political and economic change. It is weak at structural or theoretical analysis. To be fair to Davis, though, he did not write this book to impress the academic community. It is a book to be recommended to those readers interested in the underside of international relations and the politics of elite domination.

         GARY HAWES
University of Michigan
Ann Arbor

HAVENS, THOMAS R. H. *Fire across the Sea: The Vietnam War and Japan 1965-1975.* Pp. ix, 329. Princeton, NJ: Princeton University Press, 1987. $37.50. Paperbound, $17.50.

What did the war in Vietnam reveal about Japan? This is the question raised by Thomas R. H. Havens in his study of the official and the public response in Japan to the Vietnam war. He concludes that to the Japanese it was "a fire across the sea," an episode that did not need to involve their nation deeply. Nonetheless, Havens demonstrates that the war led to significant cultural and economic changes in Japan.

Havens shows that Japan made an enormous contribution to the American war effort. American bases in Japan were used as transit points for shipments of men and supplies to Vietnam. Okinawa was used as a base for air attacks against North Vietnam. Washington contracted for vast amounts of war material, the repair of equipment, and the hospitalization of its wounded in Japan.

Meanwhile Prime Minister Satō faced increasing opposition to his Vietnam policy. The major opposition parties and much of the press criticized the war. The most creative of the new antiwar groups was the Citizens' Federation for Peace in Vietnam, or Beheiren. It was organized in 1965 by Oda Makoto, who insisted on nonviolent action and who required his followers to accept responsibility for their individual actions based "on the circumstances of the moment."

In maintaining his support for the American war, Satō could rely on the traditional deference Japanese gave to those in authority. In addition, Havens points out that Satō's efforts to restore Japanese sovereignty in Okinawa worked to blunt criticism at home for his support of America in Vietnam. Simultaneously, he pressed Washington to restore Okinawa to Japan lest the opposition in Tokyo use this issue in the Vietnam debate.

Japan made enormous profits by way of the Vietnam war. By 1966, American purchases of war material helped pull Japan out of a minor economic downturn. Havens estimates that the war allowed Japanese firms to earn at least an extra $1 billion per year on the average from 1966 to 1971.

In light of these considerations, it is no wonder that the antiwar movement was unable to change official policy. In fact, Havens may have gone too far in suggesting that the protest movement limited the government's support for the war. He is on safer ground in observing that Beheiren cleared the way for a new form of citizen involvement in politics.

While Havens has attempted to include too much divergent material in one volume, this fault is counterbalanced by his sturdy and judicious judgment on major issues. This book clearly will become the standard on the subject.

EDMUND S. WEHRLE
University of Connecticut
Storrs

KARLSSON, SVANTE. *Oil and the World Order: American Foreign Oil Policy*. Pp. 308. Totowa, NJ: Barnes & Noble Books, 1986. $29.50.

KUPCHAN, CHARLES A. *The Persian Gulf and the West: The Dilemmas of Security*. Pp. xiv, 254. Boston: Allen & Unwin, 1987. $39.95. Paperbound, $14.95.

Oil and the World Order traces U.S. oil policy from before World War I under a "British-organised world order," through two post-World War II decades, when the United States controlled the world order—mainly through "control of the international oil market"—to the present period, beginning in the late 1960s, when it lost control of both. It is difficult to know exactly what was lost, however, because the criteria of world-order control are never specified.

Karlsson, who teaches peace and conflict research at the University of Göteborg in Sweden, presents a great deal of information about world oil, although few of his citations are dated in the 1980s. He also has language problems, as do his editors. There are numerous instances of awkward, confusing, and incorrect use of words, such as "as soon as the shortage became abundant again."

The major problem with the book, however, is the large number of questionable assertions. A sample follows. Karlsson states in the introduction that the United States actually wanted increases in the price of Middle East oil in the early 1970s because competitors in world trade were paying less than the U.S. domestic price and because richer Arabs would more readily compromise with Israel; the two points, however, are never again mentioned, let alone substantiated. He tells the reader that Japan dominates the world computer industry and that the Trilateral Commission became the international manager when global cooperation replaced U.S. hegemony.

Completely ignoring the Muslim rebellion against Christian President Chamoun, Karlsson indicates that U.S. troops landed in Lebanon in 1958 "to prevent the new Iraqi government from nationalising" the Western-owned Iraq Petroleum Company. He also confuses Shiites demonstrating in the east of Saudi Arabia in December 1979 with primarily Sunni, anti-Shiite, Wahhabi fundamentalists who seized the Grand Mosque in the western city of Mecca in November, moving Mecca to the east in the process. And so it goes. Karlsson's lack of command regarding such materials leaves one wondering about his detailed analyses of oil policy.

The Persian Gulf and the West is an altogether different book. Kupchan's use of language is impeccable. His judgments and conclusions are reasonable and supported by evidence. He is concerned with U.S. security policy in the Third World and focuses upon the Persian Gulf as a case study in regional policymaking. He finds three major security dilemmas.

The first is strategy versus capability. U.S. strategy has been to secure Western access to Gulf oil by containing the Soviet Union and maintaining political stability in the Gulf states. When the Soviets invaded Afghanistan in 1979, however, the regional power that had been armed to achieve these objectives, the shah's Iran, had already collapsed in revolution. No capability existed, therefore, to deter the Soviets, just as the United States had been incapable of preventing the shah's overthrow. The U.S. response toward the Soviets was reasonable in a military sense: the creation of a five- to six-division Rapid Deployment Force (RDF), a credible deterrent in that "a Soviet invasion [of Iran] would clearly be a costly and risky undertaking."

The second dilemma is globalism versus regionalism. The RDF was conceived with the global view of containing the Soviets. But the states of the region were more concerned with their own problems. The RDF could not be deployed without extensive infrastructure within the Gulf states. Heavy involvement with the United States was politically risky for them, however, because of U.S. aid to Israel and an Arab culture hostile to Western imperialists. In addition, the RDF was developed primarily to fight conventionally with the Soviets when it would more likely be called into use to deal with a regional problem, such as a coup in Saudi Arabia—a mission for which the force was not trained and equipped.

The third dilemma is unilateralism versus collectivism. Although Gulf oil is of greatest importance to Western Europe, the United States was unable to get its allies to contribute to the RDF in a substantive way, such as compensating with their own troops for U.S. forces diverted from Europe to the Gulf. "The problem was that each had a differing conception of the benefits to be derived from a collective stance [and the] United States wanted the autonomy of unilateral action."

Among policy implications produced by Kupchan's careful analysis are the following: drawing regional experts more deeply into decision making to compensate for a bureaucratic preoccupation with the Soviet threat; expecting "less overt cooperation and less consistency from the Saudis and other conservative states"; enhancing "the RDF's ability to address [non-Soviet] contingencies"; and seeking "strategic cooperation between the allies... on a multilateral, not a collective basis." In timely fashion, he comes down particularly hard on the idea of U.S. escorts for tankers in the Gulf and places greatest emphasis on reducing the West's dependence on oil from that area. These are, indeed, two fine policy recommendations.

RICHARD J. WILLEY

Vassar College
Poughkeepsie
New York

MALIK, HAFEZ, ed. *Soviet-American Relations with Pakistan, Iran and Afghanistan.* Pp. xiii, 431. New York: St. Martin's Press, 1987. $37.50.

BENNIGSEN, ALEXANDRE and S. ENDERS WIMBUSH. *Muslims of the Soviet Empire.* Pp. xvi, 294. Bloomington: Indiana University Press, 1986. $29.95.

The papers included in Hafez Malik's volume were originally presented at a seminar organized in December of 1984 by Villanova University. Malik's introductory essay outlines the basic concepts of superpower attitudes toward small states, their strategic interests, and their military interventions. The introduction sets the stage for the examination of the major themes of the three-dimensional analysis of American and Soviet relations with Pakistan, Iran, and Afghanistan.

Throughout the collection, Pakistan emerges as the key regional element of this relationship. "It is to the Soviets what Cuba is to the US, " says Morris McCain. "Sandwiched between India and the USSR, Pakistan faces a precarious future much as Poland did in the 1930s when it was hemmed by Germany and the USSR," observes Malik. Agha Shahi, a former minister of foreign affairs of Pakistan, believes that "to make Pakistan's nuclear program a centerpiece of the new US-Pakistan relationship would be contrary to the interests of both states." He also wonders why Pakistan was singled out when India's and Israel's nuclear programs have not been very strongly opposed by the United States. Lawrence Ziring believes that "the Soviets seriously contemplate the reorganization of the geography of northern Afghanistan." He also sees the dismembering of Iran and Pakistan to be "within the range of probability."

Although invited, Yuri V. Gankovski, of the Institute of Oriental Studies in Moscow, did not attend the conference, but the paper on Soviet-Pakistani relations that was coauthored by three of his colleagues was added to the volume. In spite of its obvious propaganda, the article surveys a number of Soviet initiatives pertaining to the region: nonproliferation of North Atlantic Treaty Organization and Warsaw Pact activity in Asia; the proposal for a peace zone in the Persian Gulf; reducing military activities in the Indian and Pacific oceans; and so forth.

The essays on Iran do not bring many new insights. They are mostly dominated by historical parallels and overviews. Perhaps the most interesting is Richard Cottam's analysis pointing out that the superpowers have lost control of regional dynamics: both the Russians and the Americans are at the same time allied to the Khomeini regime and opposed to it. Cottam concludes that Iran could establish a working relationship with either superpower.

Continuous stalemate in Afghanistan raises the question of the country's future. Speculations as to whether or not the Sovietization of Afghanistan would eventually succeed lead to more thorough examination of related issues. This can be seen in Henri Bradsher's account of the formation of the Communist movement in Afghanistan, as well as Louis Dupree's analysis of Afghan cultural reaction to Soviet invasion and Nancy Dupree's examination of Afghan refugees in Pakistan.

The quality of the papers and the discussion, synthesized in the last chapter, reflect strong interest and expertise, though not always new insights. The organizers of the seminar should be particularly commended for securing the participation in the discussion of so many first-class specialists from both government and academia and also for their attempts at bringing in Soviet participants.

The latest book coauthored by A. Bennigsen and S. E. Wimbush provides some of the most basic information for the study of Soviet Muslims. Its first part gives a brief overview of the spread of Islam in Russia, Islamic practices in the Soviet Union, official Islam, and demographic and other statistical data. The second part provides basic information on the history, languages, culture, demography, and politics of Muslim ethnic groups in the regions of Central Asia, Transcaucasia and North Caucasia, European Russia, and Siberia.

The book reflects Bennigsen and Wimbush's thorough familiarity with Soviet sources. The text is supported by rich demographic information arranged in over eighty tables. The bibliography is very extensive, listing the works in Russian as well as those in other languages.

By bringing together so much material in such compact format, Bennigsen and Wimbush created an excellent reference that should be welcomed by all students of Soviet Islam.

MICHAEL LENKER
University of Pennsylvania
Philadelphia

PASTOR, ROBERT A. *Condemned to Repetition: The United States and Nicaragua.* Pp. xvi, 392. Princeton, NJ: Princeton University Press, 1987. $24.95.

This book could easily have been called "A History of Nicaragua and United States Political Relationships." *Condemned to Repetition* traces the origins and development of Nicaragua from 1522, when the Spanish arrived, to its break with colonialism in 1821, and it pinpoints the U.S. entry into the Caribbean when the United States joined forces with Nicaragua in expelling Great Britain from the Mosquito Coast. After the United States decided to build the Panama Canal, the relationship with Nicaragua deteriorated. But U.S. concern with Nicaragua continued because Nicaragua "offered the only alternative canal route to that of Panama."

Pastor stated that he was striving to supply the reader with a balanced account of this subject as he delved into the policy and political machinations of the Carter and Reagan administrations concerning Nicaragua. My concentration on this was intensified when Pastor stated that he was a member of the Carter policy team. I do not feel qualified, even after looking more closely, to make a judgment about the issue of balance. Consequently, I will leave that decision to the readers.

But I can state with confidence that if readers are remotely puzzled about why we relate to Nicaragua, over and above the reasons stated in the Monroe Doctrine, they will receive clarity from this book. It is a relationship—on a deeper level of abstraction—born out of mutual political benefits, needs, frustration, mangled political and personal communications, and decision making. It is a marriage, albeit out of convenience.

The book chronicles the transactions of the Roosevelts, Hoover, Kennedy, Carter, Reagan, and others. There is another important contribution of this book, other than the history and in-depth analysis and evaluations of the Carter and Reagan policies, postures, and attitudes toward Nicaragua. It is a recommendation focusing on what the United States should do in the future as it develops Latin American policy in general and Nicaraguan policy in particular.

Pastor recommends, on page 313 under the heading "A Positive Mutual Fulfilling Prophecy," that a new thinking process be introduced, suggesting, by implication, that the combative either-or Aristotelian model of thinking be supplemented by a newer mental model. My suggestion is that the new thinking process be along the lines of Kenneth Boulding's mental models or Gaston Bachelard's fourth stage of the epistemological profile, which embraces Korzybski's and Bois's methodologies. The process would thereby enable the policy participants to shift from linear to holistic thought patterns.

When I read a book for review, I search for an author's inquiry method. The major process missing from this book is an explicit revealing of the thought patterns prompting Pastor to suggest this holistic line of thinking. From that explicit focus would emerge how we should proceed to educate our public officials to operationalize a proper process. Pastor's recommendations presuppose that with a little thought, desire, and goodwill this paradigm shift will come about. On the contrary, such a shift requires that we not only suggest this nonlinear posture but identify and design methods for the policymakers. The book does not do this. In that way, it is like most of the books that merely describe, diagnose, and forecast the outcome of a situation without developing vehicles and methodologies of an epistemological nature that would enable us to accomplish the objectives of an innovative policy. Such a methodological approach would emphasize a thinking process designed to change the very structure of thinking. This would be a logical sequence for this book to cover.

In my judgment, the book is still a stage above most and should be read universally. It is a significant contribution to our understanding of a contemporary complicated and essential phase of our foreign policy. It is well written, and other authors writing on such intricate and complex issues could benefit from taking note of the organization, clarity, and flow of this book.

WILLIAM J. WILLIAMS
University of Southern California
Los Angeles

RUDOLPH, LLOYD I. and SUSANNE HOEBER RUDOLPH. *In Pursuit of Lakshmi: The Political Economy of the Indian State.* Pp. xvii, 529. Chicago: University of Chicago Press, 1987. No price.

This is an excellent and valuable book, one that brings years of experience in studying India to bear on the subject of the state and its dominating role in that country. It is, however, designed not only as a statement of the Rudolphs' views on major issues but also as a textbook. Here it will fail. It is too narrow for use by undergraduates in a general course on South Asia, as it does not fulfill the need for a survey of the vast historical, political, social, and economic data on India. It appears best suited either for a restricted graduate-level course on the political economy of India—as the title indicates—or as a supplemental text. After well-written parts on the state and politics, the style bogs down in the final two parts on the economy and demand politics. The documentation is extraordinary; there are 89 pages of notes, many of them citing several sources. Indeed, everything available on each topic seems to have been read.

The Rudolphs maintain that India is and will remain a centrist state, one in which class politics has not played and will not play a major role. Class politics assumes two actors, private capital and organized labor. In India, however, with a highly regulated private sector—and much state investment in private corporations through such agencies as the Life Insurance Corporation—and much of labor in the unorganized sector of the economy, there is a third actor, the state itself. This third actor is centrist in politics and dominates the economy.

There is a useful theoretical discussion of demand politics associated often with liberal democracies and command politics seen most frequently in authoritarian regimes. The Rudolphs argue that the pace of economic development in India has varied little whether liberal or authoritarian governments are in power. Three demand groups are discussed in detail: labor, students, and agriculturists. In the last group, the rise of what the Rudolphs call "bullock capitalists" is the key phenomenon of the 1980s. Bullock capitalists are, generally, those self-cultivators with 2.5 to 15.0 acres of land. Their demands include lower costs for inputs and higher prices for outputs. They are at the root of India's self-sufficiency in food and are mobilized to express their views in voting and in demonstrations and other extraparliamentary means. The discussion of their role is thorough, even if somewhat long and drawn out.

This is a book that will be on the must-read list for anyone dealing with South Asia. Many of the concepts can be equally well applied to the other three major countries in the region: Pakistan, Bangladesh, and Sri Lanka. Its value to comparative politics as a subdiscipline is similarly high.

CRAIG BAXTER
Juniata College
Huntingdon
Pennsylvania

EUROPE

BRETTELL, CAROLINE B. *Men Who Migrate, Women Who Wait: Population and History in a Portuguese Parish.* Pp.

xv, 329. Princeton, NJ: Princeton University Press, 1986. $39.95.

The underlying purpose of this book, according to Caroline B. Brettell, the author and project director, Family and Community History Center, Newberry Library in Chicago, is "to explore the relationship between emigration and a range of other demographic phenomena (including fertility) in a local context and, in doing so, to examine the connection between population patterns and sex roles." I believe Brettell has satisfactorily attained her stated aim through this well-researched and thorough study.

Brettell conducted her fieldwork in Lanheses, a parish in northern Portugal that has seen the largest rural out-migration since the end of the eighteenth century. She came to know this community through emigrants to France, the site of her earlier field research.

Brettell's methodology is ingenious. She combined ethnographic fieldwork and oral histories with demographic analysis of parish registers and census material, emphasizing the late nineteenth century and the twentieth century.

She concludes that "emigration from northwestern Portugal is closely related (though not always causally) to high ages at marriage, high female celibacy, lowered fertility, and high illegitimacy." She further explains that "it is the unfavorable sex ratio—an oversupply of women—resulting from male-biased emigration which has influenced both sexual behavior and sexual mores." Brettell cautions, however, that the cultural context of Portugal should be considered in explaining the sex ratio—factors such as "the distribution of political and economic power, the degree of openness or fluidity in society, and the balance between the sacred and the secular" should not be ignored. All of these factors, Brettell maintains, will influence the relationship between men and women; for instance, the extent to which the women's virginity is highly valued in society affects the liberty and self-determination that women can expect. Finally, Brettell states that the roles of women in northwestern Portugal and the coastal regions of the province of Galicia, Spain, are somewhat similar, a significant finding in the comparative study of European ethnography and emigration.

Brettell's book has both weaknesses and strengths. Nonspecialists, for example, may find the voluminous quantitative data intimidating. The book seems repetitious in some parts. It needs more photographic illustrations. On the other hand, Brettell has fruitfully blended the methods of demography, history, and anthropology in the analysis of emigration. It is a solid contribution to anthropological methodology and especially to European ethnography. Her research will undoubtedly serve as a model for comparative studies of emigration worldwide.

Brettell and Princeton University Press deserve our gratitude and praise for publishing this volume, destined to be a classic in the field.

MARIO D. ZAMORA

College of William and Mary
Williamsburg
Virginia

GORDON, LINCOLN with J. F. BROWN, PIERRE HASSNER, JOSEF JOFFE, and EDWINA MORETON. *Eroding Empire: Western Relations with Eastern Europe.* Pp. xv, 359. Washington, DC: Brookings Institution, 1987. $31.95. Paperbound, $11.95.

The Soviet Union's empire in Eastern Europe has never met its expectations, implied or expressed. In an extreme sense, it was considered a buffer against anti-Soviet attack, but such attack has never been contemplated seriously except by a few interdictionists in the United States. More evident, it was a region open to economic exploitation; this purpose was met more fully, but there have been manifest shortfalls. It has been a subject for replication of the Soviet Union's domestic political system and its ideology; but replication has never occurred except under conditions of such political illegitimacy that heavy-handed crises have occurred. Least, it has been a target of cultural influence, but it has been unlikely for this influence to take root in regions so diverse and so rich in their own histories.

All of these expectations have been played out against a backdrop of Western influence. In a period of *glasnost*, accelerated change within the empire may well occur. This book examines such a possibility and offers both analysis and caveats for the policy-oriented public. Its principal writer appears to consider it axiomatic that the West has the right, and even the duty, to motivate this change; it is not evident that all the authors share this view.

This volume is rooted in the Brookings Institution's historic interest in U.S. public affairs and is the result of a 1985 conference on the future of Western policies toward Eastern Europe. It was designed by Lincoln Gordon. The volume's chapters are updated to mid-1987.

Following a general introduction by Gordon stating the design of the study and overview of findings, Brown briefly describes the East European setting and its relationships with the West. Gordon then begins the longer chapters that give views from Western countries with his chapter on the perspective from Washington. In the same vein, Joffe writes on Bonn, Hassner on Paris, Moreton on London, and Brown on Vienna and Rome. Gordon concludes with "Convergence and Conflict: Lessons for the West." The volume's detailed focus on trade by the Eastern bloc with the West is implemented by 14 tables.

Given the central themes of change and erosion within the East—absent Albania and Yugoslavia, which are defined out of the group of countries discussed—each operative chapter addresses three alternative positions available to the West: accommodation, transformation, and dissolution. The first, at an extreme, would accept Soviet control of the individual Eastern countries with no attempt to stimulate change. Dissolution, at an extreme, would seek to overthrow a Soviet link by all possible means. Transformation "plus" is suggested as characteristic of U.S. policy—a policy assuming that reforming change should be urged and at times even extorted—and transformation "minus" as that of the Federal Republic of Germany, given its long-term, always frustrated goal of reunification. These two have the greatest concern and most active policies. The other Western countries generally take minus positions as well, but less active ones. All six vary over time as much as do the Eastern six. In addition, each chapter touches on the individual Western countries' roles and attitudes toward international entities—the North Atlantic Treaty Organization, International Monetary Fund, International Bank for Reconstruction and Development, European Economic Community, and Organization for Economic Cooperation and Development—and on recent European confidence building and balanced-force conferences.

The collection is, like all those resulting from symposia, somewhat uneven. Gordon's chapters are both the longest and the most insightful, to be expected because of his long experience with U.S. foreign policy toward Europe and because of his formative scholarly role in the project. He explains that his materials included extended interviews with currently serving officials. Chapters by Joffe, Hassner, and Moreton are careful and adequate, the work of experienced political scientists. The chapters by Brown unfortunately lack the insight and care of the others. Joffe deals intensely with the subtleties of German reality, in which West and East Germany both converge and conflict. Given the context of Austrian and Italian interests and roles, Brown could have been equally sharply definitive and provocative, but he is not. On the whole, Hassner and Moreton write that neither France nor Britain consistently has broadly developed nor sustained interests in the six Eastern countries; rather, their policies are conceived and executed generally against the backdrop of the imperium.

In some respects, the book is a compromise, reflecting the constraints of publication space as against the writers' wish to project their insights of a complex relationship. As such, it is more an introduction for an already partially informed audience than a text for policymakers. Gordon concludes on a note of realism as well as hope. The West can have ever-increasing effect on the East, but to do so demands clarity of vision and a patient understanding that changes come slowly and reforms almost glacially. At no point is the axiom—of the rectitude of the West's effort to create change in the Eastern bloc countries—examined critically.

     PHILIP B. TAYLOR, Jr.
University of Houston
Texas

KOONZ, CLAUDIA. *Mothers in the Fatherland: Women, the Family, and Nazi Politics.* Pp. xxxv, 556. New York: St. Martin's Press, 1986. $25.00.

Claudia Koonz's wide-ranging effort serves to highlight the strengths and some of the weaknesses of contemporary gender analysis, here in the form of social history. Before we had the locution "gender analysis," several important discussions of women and the family in the Third Reich had appeared, including Clifford Kirkpatrick's early, pathbreaking, *Nazi Germany: Its Women and Its Family Life* (1938). More recently, Leila J. Rupp's *Mobilizing Women for War* (1978) and Jill Stephenson's *Nazi Organization of Women* (1981) helped to pave the way for Koonz's more exhaustive treatment of this complex and morally fraught subject.

Controversially, Koonz insists that those Nazi women who saw themselves as leaders of a freedom movement and tagged themselves feminist should, indeed, be taken seriously rather than denied any connection with terms we prefer to sanitize or to celebrate. What kind of feminism was this— seeking an honored role for women inside the Nazi party and state? What were the terms in and through which women's liberation was construed? Koonz distinguishes the actions and reactions of Protestant women, who tended, with their male counterparts, to be more solidly in the pro-Nazi bloc, from those of Catholic women, whose Church ties and loyalties distanced them from thorough immersion in Nazi blandishments, especially where Nazi eugenics and biological hygiene programs were concerned. Neither group comes off very well, however, in their relations with Jewish women. Just a few, according to Koonz, acted on their empathy for the suffering of their Jewish counterparts. Worst of all, many apparently shut this suffering out completely or sanctioned it tacitly or openly.

On this and other matters, however, some of the drawbacks as well as strengths of gender analysis emerge. For example, Koonz argues, "Jews, after all, and Catholics, to a lesser extent, recognized Hitler's hostility toward them and avoided his movement. Why did women seem not to notice?" This statement strikes the reader as strange; women, after all, are at least one-half of the Jewish and Catholic population. In another case, Koonz makes repeated statements about the motivations of Nazi women, about their inner identities, about their ability to screen out misogyny in Christianity, making it easier for them to do the same with Nazism. Yet she insists that statements by Nazi women are "exceedingly rare." What form of epistemological privilege is at work that enables her, as a researcher, to ferret out psychological imperatives about which her subjects themselves were, or are, silent? A claim such as "Nazi women accepted the promise of second-sex membership in Hitler's movement in exchange for the hope of preserving their own womanly realm against male interference" requires a good bit more than circumstantial evidence to buttress.

Mothers in the Fatherland suffers from excessive use of simile to carry the argument. In chapter 3, for example, I counted five similes, each aimed at drawing connections between Christ and Hitler; the Nazi movement and a religious crusade; church altar-care societies and Nazi women's work; a Nazi women's leader and Hitler; Hitler and a matinee idol. The connections become strained and the reader begins to ponder whether so many "likes" are a substitute for solid substantiation about the nature of Nazism, Nazi women, and Hitler.

These rather striking weaknesses in the nature and form of argumentation and documentation aside, Koonz's book is must reading for those interested in the Nazi era and in women's history. She helps to pinpoint a series of conundrums that haunt all feminist movements everywhere, not just in the Third Reich. Should women insist on a parity of power with men within coequal spheres of public and private life, or are they best advised to sustain and to make the most of those realms within which the female writ has long run, family and community life? Koonz pretty much blasts Nazi feminists for celebrating and seeking to control absolutely a separate female realm. Yet she herself refers to something similar as a source of particular gender-based moral values when she insists that women guards in Nazi camps has a more difficult time being brutal than did the men. She writes: "For a woman to become a guard required so major a departure from the normal values and experiences of women, perhaps the few who ended up on camp assignments were more apt to be depraved or deranged than the men." Whether or not this is in fact the case, I cannot say, but it does suggest that for Koonz, and many others, there is something to be said for "the normal values and experiences of women."

JEAN BETHKE ELSHTAIN
University of Massachusetts
Amherst

ROSE, RICHARD. *Ministers and Ministries: A Functional Analysis.* Pp. ix, 287. New York: Oxford University Press, Clarendon Press, 1987. $52.00.

This useful inquiry into the functions of government ministries originated in papers presented at an annual conference of the Political Studies Association Work Group on United Kingdom Politics. Each of the papers—by Richard Parry on the Scottish Office, Ian C. Thomas on the Welsh Office, and P. N. Bell on Northern Ireland—was about a ministry its author knew from working in the civil service. Revised, they have been augmented by a series of chapters by Richard Rose exploring the similarities and differences between Whitehall ministries. While the chapters on territorial ministries provide informative descriptions of their organization and functioning, it is the broader sweep of Richard Rose's general observations that will be of interest to most students of British history and politics.

In his introductory sections, Rose makes clear that the purposes of ministers and ministries are customarily not the same. For the minister, who is answerable for the work of a ministry but very infrequently manages it, the organization serves as a vehicle for promoting political ambitions. To say that a minister is a professional politician before he—it has almost always been "he"—is a secretary of state is simply an accurate description of personal priorities and of ministerial careers. There is an organizational succession among the ministries— among equal bodies some are considerably more equal than others and few ministers want to stand still once they have arrived in the cabinet. It is the ministry, not the minister, which in reality "carries on the Queen's government." Noting the inertia of public organizations that is a common feature of government, Rose comments that within a limited political unit of time—say, the life of a Parliament—such inertia is much stronger than the scope for any organizational change. Whatever changes may be publicly announced, usually almost every program of a ministry continues as before. As for the allocation of resources within the ministry, the secretary of state, though nominally responsible, presides over a divided organization in which most of the disbursement of funds is carried on by personnel—in local government, nationalized industries, the health service, for example—not directly answerable to the ministry, let alone the minister.

In summarizing his findings, Rose argues

that the managerial reforms the Thatcher administration has been promoting in recent years are based on the incorrect assumption that, as in the business world, individual career incentives and resource management are mutually reinforcing. At best, he concludes, there is a random relationship between political status and program resources. "At the centre of government, the necessities of politics matter more than managing and spending big sums of money." The case, confirmed, if sometimes indirectly, by the three territorial studies, is a persuasive one, resting on a substantial base of solid research illuminated by the good sense we have come to expect of Richard Rose.

     HENRY R. WINKLER
University of Cincinnati
Ohio

SYMONDS, RICHARD. *Oxford and Empire: The Last Lost Cause?* Pp. xviii, 366. New York: St. Martin's Press, 1986. $29.95.

Richard Symonds, a member of St. Anthony's College, has little difficulty in convincing the reader, from the outset of this delightful study, that Oxford University did play a preeminent role in British academe in the propagation of the imperial mission. The university fed its graduates into the Indian Civil Service, where it enjoyed a near monopoly; it nurtured such "proconsuls of empire" as Curzon and Milner, and absorbed the stream of bright and athletic young men sent to it from the colonies, the Commonwealth, and the United States due to the beneficence of the Rhodes Trust. Yet Symonds's chosen subtitle suggests that Oxford's impact was, at best, an ambiguous factor in the rise and fall of the British Empire.

The radical politician John Bright once misrepresented Oxford as "a home of dead languages and undying prejudices." Symonds rejects this slight, having a more interesting and subtle tale to tell of the impact of Oxford's classical education and inbred tradition of confidence upon empire building.

Aristotle, Plato, and Thucydides were all carried in the knapsacks of young imperialists educated at Oxford. The classical philosophers inspired rules of political conduct and moral principles and, incidentally, confirmed the fitness to govern of a small intellectual elite. Thucydides and the historians of Greece and Rome provided a ready compendium of historical metaphors for action and handbooks of strategic maxims often, one suspects, misapplied. Milner's famous "Helot" dispatch from South Africa describing the grievances of the Uitlanders is one such false analogy pointed out by Symonds. Unfortunately, in the broad sweep of his study, which ranges from 1870 to 1939, Symonds has relatively little space to devote to any detailed study of the influence of the classical imagination upon the handling of empire. Indeed, it seems clear that classical images of empire overlapped in this period with the new language of social Darwinism. Just how these two readings of empire coexisted would make a fascinating study.

If Oxford's classical education—particularly the degree known as "Greats" for "Classical Moderations and Greats"—provided in some ways a handbook for empire, Oxford's vigorous defense of its supreme value in the curriculum also proved one of the strongest factors in inhibiting Oxford's drive to become the great imperial university. Time and time again in Symonds's book, one reads of the fears of Oxford's educators that the classical education would be eroded by demands for various kinds of utilitarian learning, particularly if Oxford embraced its imperial role too single-mindedly. This rearguard action was a lost cause, and Symonds gives some credit to the Rhodes scholars themselves for contributing to the momentum of change that led to the abolition of compulsory Greek and the opening up of graduate studies.

What about Bright's charge of "undying prejudices"? In his conclusion, Symonds records the "insensitivity" of Oxford men on the subject of race. This may be too sensitive a reading by far, for it is difficult to accept Symonds's view that the use of the term "race" in a patriotic sense "did not necessarily imply a contempt for or hostility to other races." Racial scorn was still very much a component of British grand strategy as late as the early years of World War II and contributed to grave errors in policy toward states such as Italy and Japan, known politely as second class. Turning away from this issue, Symonds does do service in pointing out that whatever were the wellsprings of racial prejudice in British culture, Oxford rarely stooped to making propaganda for empire causes. The university, Symonds concludes, offered no systematic indoctrination into the philosophy of empire or the civilizing mission of Britain to its students, whether inward bound from the empire or about to embark on imperial careers.

The favorite vision of some of Oxford's educators to turn the university into a great imperial university was, in the end, never achieved. Symonds lists many reasons why not, including the imaginative failure of the university to turn the energies of its female students, often enthusiasts of imperial causes, to best use. But Oxford did become an important conduit of values and of people and would still be so in the wider Commonwealth sense were it not for the current ruinous and shortsighted policy of discriminatory fees for foreign students imposed on the university by one of its own.

WESLEY K. WARK
University of Toronto
Ontario
Canada

UNITED STATES

CAIN, BRUCE, JOHN FEREJOHN, and MORRIS FIORINA. *The Personal Vote: Constituency Service and Electoral Independence.* Pp. x, 268. Cambridge, MA: Harvard University Press, 1987. $25.00.

The relationship between legislators and their constituents has become an increasingly important area of study. Once ignored by congressional scholars, the subject has through the work of Richard Fenno and others been energized and now is seen as central to a number of crucial topics in electoral behavior and legislative policymaking.

But despite the outpouring of work in this area over the past decade, there remain a number of unsettled controversies. What motivations do legislators have for constituency service? What is the electoral value of these activities? For example, do casework activities on the part of representatives increase their visibility, reputations, and vote-getting powers? What are the broader policy implications of constituency service for the governing capacities of political systems? And how do these activities stand up in comparative context, that is, in countries featuring different rules, institutional settings, and political context?

Bruce Cain, John Ferejohn, and Morris Fiorina have written in their book a splendid account of the who, what, where, how, and why of constituency service in the United States and Great Britain. Impish reviewers might retitle this volume "Everything You Wanted to Know about Constituency Service but Were Afraid to Ask."

Cain, Ferejohn, and Fiorina marshal an impressive array of data in this study: interviews with 102 administrative assistants in Congress and survey data from 1958, 1978, and 1980 in the United States, as well as interviews with 69 members of Parliament and 32 agents in Great Britain and British survey data from 1963, 1964, 1966, and 1979.

In chapter 9, they summarize their most important findings. Few voters consider constituency service unimportant and most are satisfied by the services they receive. Legislators who engage in extensive service activity are better known, more favorably evaluated, and more successful electorally than those who do not. Constituency service figures more prominently in evaluations of members now than in earlier periods, and the electoral payoff has grown over the past two decades.

The central contribution of this work lies in its comparative focus. There now is an extensive literature on constituency service in the United States. Little work, however, has appeared on this subject from a comparative framework. Although Cain, Ferejohn, and Fiorina find a number of similarities between Great Britain and the United States, there also are several important differences. Members of Congress adopt a less personal, staff-oriented style of constituency service, while members of Parliament adopt a more personal, time-intensive style. In addition, as expected given the mature case of constituency service in the United States and the developing case in Great Britain, there are greater electoral payoffs for American than British representatives.

Cain, Ferejohn, and Fiorina's broadest conclusions are encapsulated under the rubric of a "personal vote." They argue that successful representatives earn personalized as opposed to party-based support. This source of power accentuates the decentralization of governing institutions and increases the role of interest groups in the policymaking process. These systems furthermore tend more often than other systems toward stalemate and inconsistency.

The one topic I would have liked to have seen addressed more explicitly is prescriptions for the maladies presented by personalistic politics. Other than turning the clock back to a party-based system—which in any event does not appear very likely—it is not clear exactly what options opinion leaders have for constraining interest group influence or the electoral independence of legislators. Until these issues are addressed, the personal vote will continue to create problems for policymakers in Western democracies.

DARRELL M. WEST
Brown University
Providence
Rhode Island

COSGROVE, RICHARD A. *Our Lady the Common Law: An Anglo-American Legal Community, 1870-1930.* Pp. x, 330. New York: New York University Press, 1987. $40.00.

Eight prominent men significantly involved in legal crosscurrents between England and the United States from 1870 to 1930 are discussed in this volume. Langdell, Bryce, Holmes, and Pollock spearheaded the bilateral celebration of the common law; Maitland, Pound, Frankfurter, and Laski contributed to its demystification by advocating new political and intellectual issues. American legal realism contributed to the end of the confraternity of elite lawyers.

Langdell, whose case method revolutionized legal education at Harvard, espoused law as a science and "thinking like a lawyer," a phrase still echoed in today's law schools. Bryce promoted an elitist Anglo-Saxonism. American law was firmly grounded in the common law system but not to the extent claimed by the romance of Anglo-Saxonism. Holmes, who rested for a long time on the laurels earned by the publication of *The Common Law*, believed firmly in the doctrine of "my country right or wrong" and bluntly stated that he hated facts! Holmes led the American response to the problems posed by Austin for the common law. Through him, English legal thought had its greatest influence in the United States. Pollock, who exerted greater influence in the United States than in his native England, argued that "our lady the Common Law is not a task-mistress but a bountiful sovereign whose service is freedom."

Maitland corresponded copiously with American legal academics. Unlike Pollock, he was critical of the historical jurisprudence of Maine and Austin. When myth gave way to history, the unraveling of Anglo-American homogeneity began. Pound created a jurisprudence that owed more to American conditions than English forebears, emphasizing comparative and sociological jurisprudence. Frankfurter, an unabashed anglophile, worshipped the common law in its native setting, attempted to promote an American meritocracy, but associated himself with radical causes and plunged into politics. An advocate of self-restraint, he ultimately became a conservative, strict-constructionist jurist. Laski, who, though not a lawyer, initially personified the Anglo-American legal community, questioned such sacrosanct areas of the common law as the traditional rights of private property. By 1930, through his praise of American jurisprudence and contempt for the English version, Laski reversed the supremacy of English jurisprudence dating back to 1870.

The rise and fall of the Anglo-American community is highlighted in this almost perfectly edited book. It is scholarly, stylistically appealing, somewhat repetitious, and almost too crammed with interesting though tangential information. It is highly recommended for the jurisprudence buff, less so for the average lawyer or casual reader.

KARL H. VAN D'ELDEN

City of Minneapolis
Minnesota

Hamline University
St. Paul
Minnesota

FERMAN, BARBARA. *Governing the Ungovernable City: Political Skill, Leadership, and the Modern Mayor*. Pp. xii, 281. Philadelphia: Temple University Press, 1985. $34.95.

Governing the Ungovernable City provides case studies of mayoral leadership in two of America's most glamorous cities, Boston and San Francisco, in answer to the question, "How does a mayor acquire power in a time of dwindling resources?" Inspired by the "powerful administration" of Boston's Kevin White to "believing that strong mayoral leadership is necessary," *Governing* compares the political motives and styles of three mayors in some depth and brings in examples from the careers of others. The many themes of mayoral leadership covered include the costs of weak leadership and the costs of acquiring political power. Ferman identifies the institutional sources of mayoral political weakness, some political skills mayors use to consolidate power, and the "fine line" between acquisition of "necessary power" for legitimate programmatic ends and "self-aggrandizing power" for personal ends.

The title of *Governing the Ungovernable City* plays, of course, upon Douglas Yates's influential book, *The Ungovernable City*, but *Governing* more recalls Robert A. Dahl's seminal work, *Who Governs?* Unlike Yates, Peterson, and others who conclude that deep-rooted social problems and conflicts, together with the extremely circumscribed political authority of cities under federalism, make American cities fundamentally unmanageable, Ferman finds a "contradiction between the dim predictions for mayoral leadership in the 1970s" and the success of mayors such as Boston's Kevin White. In this, Ferman recalls Dahl's statement that political science has provided "a number of conflicting explanations for the way in which democracies can be expected to operate in the midst of inequalities in political resources [that leave] room for the politician." Ferman distinguishes *Governing* from *Who Governs?* as addressing the problems of urban leadership in an era of dwindling resources. Nevertheless, parallels to Dahl's work are instructive.

Ferman, like Dahl, uses the freedom of the case-study approach. Ferman decries the limitations of conventional methodology for understanding leadership, although the science of "outputs" and "outcomes" that she forsakes is much closer to the political science Dahl was looking for when he challenged the use of sociological methodology for explaining political phenomena. Dahl combined observation with a range of other sources and original data into a lucid and compelling exposition on political pluralism that to this day frustrates the efforts of those of us who dislike the impact *Who Governs?* has had upon political science. *Governing*, in contrast, depends heavily upon 123 interviews that Ferman often cites anonymously and unspecifically to support her explanations. This technique produces a rather unlively narrative in which the reader must accept Ferman's interpretation of, essentially, private conversations. More direct quotes—which do not have to be attributed—might have made the book livelier while giving the reader some independent capacity to evaluate the interviews. Political officials hardly qualify as expert, disinterested sources in the depiction of events and underlying motives in which they may very well have a vested interest or an ideological bias.

Ferman labors to integrate much contemporary urban politics scholarship into her thesis, a literature that has burgeoned since *Who Governs?* The first chapter builds a solid foundation for her research with the specific literature on mayoral leadership, but the book is less constructive when it comes to placing the desirability of mayoral leadership in a larger setting.

Governing is probably too specialized for students, but those with greater sophistication about urban problems will appreciate Ferman's effort to find cause for optimism. The shift of national priorities away from urban policy leaves local, self-help strategies as one of the few options. The development of mayoral leadership may prove as valuable to cities as the development of political organizing skills has to many urban constituency groups. The expectation of federal assistance trivialized city leadership and politics to the purchase of solutions. Although *Governing* trims its ambitions in its conclusions, citing the "difficulties and costs of developing that leadership in the absence of an institutional base," it is nonetheless an extremely bold challenge to conventional wisdom about the possibilities of urban politics and urban redevelopment.

BRUCE EDW. CASWELL
Temple University
Philadelphia
Pennsylvania

HIGGS, ROBERT A. *Crisis and Leviathan: Critical Episodes in the Growth of American Government.* Pp. xix, 350. New York: Oxford University Press, 1987. $24.95.

Crisis and Leviathan is written on the assumption that there was once a natural and genuinely free economy, which the mistakes of well-intentioned men, the machinations of politicians, the self-interest of a wide variety of powerful social groups, the recurrence of severe economic crises—themselves partly occasioned by human error—and the decisions of a spineless Supreme Court have worked together to destroy. Pursuing the question of how to explain such developments, Robert A. Higgs works out an elaborate methodological scheme intended to make room in economic theory for the influence of ideology and the impact of historic circumstances on national economic policies, but the main element of his book is a circumstantial account of political and economic crises during which the United States went wrong. Characteristically, he closes his volume by expressing the hope that Americans will substitute a valid ideology for the false views they have increasingly adopted, thereby enabling themselves to cope with events without the intellectual baggage that has undermined the market economy and produced an imminent threat of fascism.

Although his definition of fascism is so naive as to cast doubt on his other judgments, Higgs does depict the expansion of governmental authority over the American economy in ways that challenge the careless so-called progressive assumption that what happened was necessary and inevitable. Rather, he demonstrates that during the Progressive Era, World War I, the Great Depression, and World War II, American political leaders and interest groups took advantage of a supposed crisis and built on an interventionist ideology to bring public policies into conformity with their prejudices or even their personal advantages. Significantly, however, he measures their behaviors by the standard maintained by Grover Cleveland during the early 1890s, when, during his second term in office, he was confronted by a severe depression but stood fast against agitation for work relief, clamor for the coinage of silver, aggressive trade unionism, and the passage of a limited income tax, which Higgs views as the beginning of the end of a free economy. In other words, while he excoriates weak or willful men who distorted the American economy, his real target is history itself. Bound by the presuppositions of economic liberalism and the conventions of economic theory, he cannot imagine that reputable men would knowingly have sought to abandon the tenets or modify the consequences of a market economy. In this respect, his interest in understanding history extends no further than the hope that by explaining past errors he will be able to persuade us to undo them in the name of economic verities initially propounded by Adam Smith.

RUSH WELTER

Bennington College
Vermont

KUGLER, ISRAEL. *From Ladies to Women: The Organized Struggle for Woman's Rights in the Reconstruction Era.* Pp. xiv, 221. Westport, CT: Greenwood Press, 1987. $35.00.

The main focus of this study is on "the transformation of American ladies into women increasingly conscious of their inferior status, learning the arts of organization, experiencing the pangs of dissension within their ranks, and experimenting with political strategies." It treats the creation of permanent woman's rights organizations dedicated to securing woman's suffrage and examines the schisms, tactics, and diverse social outlooks of the leadership.

Kugler succeeds in making clear that Lucy Stone and the American Woman Suffrage Association, by combining pursuit of woman's suffrage with a "steadfast" support of the Republican Party, represented the mainstream of the woman's movement. By contrast, Elizabeth Stanton and Susan Anthony led a minority element, the National Woman Suffrage Association. Frustrated by their failure to induce longtime allies in the antislavery movement and in the Equal Rights Association to support the suffrage for both women and blacks, they exhibited "a nativist, anti-immigration bias, as well as an antiblack predilection." Victoria Woodhull, leader of the third group, is characterized as "the supreme side issue."

Kugler's treatment of the woman's movement during Reconstruction is straightforward. His overview of the era draws on C. Vann Woodward, Kenneth Stampp, and John Hope Franklin. In explaining the woman's movement, he reworks standard print materials such as Ida Husted Harper's *Life and Work of Susan B. Anthony*, the *Proceedings* of the National Woman's Rights Convention, and the *History of Woman Suffrage*, edited by Susan Anthony, E. C. Stanton, and Matilda Gage.

If Kugler offers the scholar little that is new, students will find his presentation of the main features and personalities of the woman's movement clear and orderly. He details the efforts of Anthony and Stanton to court the National Labor Union and the Knights of Labor. Kugler also explores such "side issues of contention" as marriage, divorce, and sex, as well as the woman's rights press. Here, too, if his treatment does not break new ground, he makes clear that pursuit of the suffrage was but one aspect of the movement.

Kugler's effort to place woman's rights during the Reconstruction Era in perspective is admirable but flawed. To be sure, the prewar temperance and abolition movements were training grounds for the founders of the woman's rights movement. By focusing on a handful of leaders, Kugler fails to make clear the breadth of women's participation in these prewar movements. The epilogue is marred by minor errors: Woodrow Wilson was president from 1913 to 1921; the first congresswoman was Jeannette Rankin; Carrie Catt was Mrs. Catt. His index, however, is both useful and accurate.

ROBERT L. DANIEL

Ohio University
Athens

LEWIS, JOHN S. and RUTH A. LEWIS. *Space Resources: Breaking the Bonds of Earth.* Pp. xiii, 418. New York: Columbia University Press, 1987. $30.00.

John Lewis, a professor of planetary science at the University of Arizona, and his wife, Ruth, a freelance science writer, have written an interesting and informative summary of the history of space exploration as well as problems and possibilities associated with the exploration of space and the development of space resources. The book is written to be appreciated by the layperson and will be enjoyed by anyone who is interested in space.

Divided into 14 chapters, *Space Resources* begins by presenting the history of space races and space exploration. The first chapter provides an introduction, and chapters 2 and 3, a historical summary of space races, including a description of the assumptions and planning that have motivated major space-related decisions in both the United States and the USSR up to the present time. Chapter 4 provides an interesting summary of projections made by space enthusiasts in late 1950s as to where we would be in the 1980s and evaluates the accuracy and assumptions of those projections. In chapter 5, as well as throughout many of the remaining chapters, the Lewises examine current, and the potential for future, developments in technology that will allow for increased space resource development. Chapters 6 through 11 deal primarily with an evaluation of the moon, asteroids, Mars, and Phobos and Deimos—two small satellites of Mars—and the possibilities for resource development of these bodies including what resources we might benefit from, how they may be exploited, and the technical difficulties involved in their exploitation. Chapter 12 examines the potential relationship between space resources and space travel. The Lewises describe new technologies that may be available after the development of resources in space, and the applications of those technologies. Chapter 13 provides a summary of the plans and goals for space development on the part of the countries that are involved in space exploration. Finally, in chapter 14, the Lewises present their conclusions and recommendations. Essentially, their recommendations focus on the reorganization and redirection of the National Aeronautics and Space Administration (NASA) consistent with its original mandate with greater diversification of its efforts; accelerated space research; accelerated space resource research and discovery; greater international cooperation and collaboration in space research and activity; and increased Strategic Defense Initiative research undertaken with full disclosure to the Soviet Union to enhance, it is hoped, national security.

This is an excellent and readable book. The Lewises' personal experiences add to and enliven the narration. Humor, as often as is possible, is put to good use. For example, in describing the utility of asteroids, the Lewises' intentional double entendre refers to "the impact of near earth asteroids," and in describing the possible makeup of a homestead on Mars, they make the observation that "the secret of living with goats is the same as creating successful nuclear fusion reactors: perfect containment."

Although not one of the book's stated purposes, a little more analysis of the politics of space exploration would have been welcome. The Lewises do devote several pages to an analysis of NASA politics and the incentives that agency has had in developing and pursuing big-budget, and visible, programs. For example, after discussing NASA's problems and policies in relation to future space stations and summarizing evidence presented from various sources that a space station is unnecessary, they conclude, "Why is NASA really building a space station? The answer is simple: NASA must at all times have one big project that qualifies as a National Goal. If they didn't, the vultures that patrol Old Foggy Bottom would tear it limb from limb."

Space Resources will be enjoyed by anyone fascinated with space exploration.

ZACHARY A. SMITH
University of Hawaii
Hilo

LOWENSTEIN, SHARON R. *Token Refuge: The Story of the Jewish Refugee Shelter* at Oswego, 1944-1946. Pp. vii, 244. Bloomington: Indiana University Press, 1986. $27.50.

In June of 1944, President Roosevelt established a shelter for refugees, mostly Jewish, from Nazi-occupied Europe, admitting just under 1000 persons. It was the only admission to the United States of refugees outside the quota system established by immigration laws, and it was expressly stipulated that admission to the shelter did not constitute immigration but only a temporary haven until the end of the war.

The shelter was located at Fort Ontario, outside Oswego, New York, an old Army post that had been used, off and on, for various purposes. The refugees selected were not really persons rescued or evacuated from Nazi or Nazi satellite regions. They had in fact managed to survive, and they had been living in Allied-occupied southern Italy, albeit mostly under circumstances of deprivation and various degrees of physical and certainly psychological suffering. To remove them to a place in which they could initially obtain rest, decent food, and medical attention was helpful, but it cannot be said that it was an act of life saving. Nor can it be said that it constituted more than a token effort to cope with a problem of vast magnitude.

At least as early as 1942, compelling evidence had come out of Europe that Nazi Germany was embarked upon a new and monstrous policy regarding the Jews, a policy not merely of expulsion and resettlement—although initially it was camouflaged as such—but a policy of mass murder, employing increasingly sophisticated technology and gaining correspondingly in momentum. As early as 1938, after anti-Semitic excesses in Germany and in annexed Austria made it clear that Jews would have to leave the Nazi *Reich*, some efforts were made to offer rescue and relief, offers that failed to materialize because none of the countries convened, including the United States, wished to open their doors to significant numbers—or to any numbers—of Jews.

Sharon Lowenstein chronicles the bureaucratic gyrations leading to the Oswego Shelter decision, unfolding a story that reads like a creation by Kafka or Orwell. President Roosevelt showed personal sympathy but opposed actual efforts at mass rescue for pragmatic reasons: a fear of political backlash caused by anti-immigration and anti-Semitic forces. He was further hindered by bureaucratic opposition centered in the Department of State. Thus the stringent stipulations regarding immigration were not relaxed. The net result was that only a fraction of the immigration quotas available were ever filled during all the years of World War II.

In addition to the infighting among U.S. bureaucracies, Lowenstein deals with infighting among Jewish organizations and with public opinion pressures pro and con. Most interesting is her detailed discussion of the physical aspects of shelter life, camp organization, problems of morale, of work, and schooling for children. As a bright aspect, several camp administrators and indeed a few bureaucrats emerge as persons of great commitment and compassion. While the political-bureaucratic background has been analyzed in other works, the strength of Lowenstein's study lies in the accounts of individual records, much of it based on personal interviews with former shelter residents and their families. Thus policies and statistics are reduced to the human level. *Token Refuge* leaves a bitter aftertaste at the thought of what might have been accomplished by a sincere and major effort at rescue.

HANS SEGAL

Cleveland State University
Ohio

PETERSON, PAUL, BARRY RABE, and KENNETH WONG. *When Federalism Works*. Pp. xvi, 245. Washington, DC: Brookings Institution, 1986. $28.95. Paperbound, $10.95.

As many observers have noted, discussions of American federalism have been plagued by excessive generalization on the one hand and excessive particularism on the other. Arguments about how well federal programs have or have not worked are made either in terms of the broadest generalities with few, if any, empirical referents, or in terms of particular programs in particular localities, with little attention to the generalizability of results. There have been too few efforts of the sort made in this book—the examination of multiple programs in multiple sites to produce defensible generalizations about what types of federal program efforts are and are not successful under what types of local political and financial circumstances.

Peterson, Rabe, and Wong present a vigorous and carefully argued political version of an argument frequently made by economists—state and local governments do better at running so-called developmental programs to improve local infrastructure and enhance an area's attractiveness to potential employers and residents than with redistributive social programs that attempt to address the problems of lower-income groups. Developmental efforts are popular with both local politicians and the public, and federal programs in support of these efforts can proceed with limited oversight and little controversy because they largely finance activities that locals would be trying to do in any case. Redistributive programs, by contrast, are less popular, more difficult to implement, and require more in the way of federal supervision to operate effectively. Left to their own devices, state and local governments will underspend on these types of programs; unsupervised, they will attempt to divert funds from nominally redistributive federal programs to more popular developmental purposes or attempt to avoid compliance with politically controversial program objectives. Federal regulations to prevent diversion and ensure compliance must of necessity, therefore, be more intrusive and complex than those for developmental programs. Redistributive programs can be operated effectively, but largely only where program professionals can insulate themselves from the pressures of elected officials. In cities or program areas where professional control cannot be established and redistributive programs become politicized, conflict and ineffective performance have resulted. This argument is illustrated by case studies of nominally developmental and redistributive programs in education, health, and housing and community development in four large cities—Baltimore, Milwaukee, Dade County, and San Diego—that illustrate a wide range of economic, political, and financial circumstances.

The policy conclusions Peterson and his colleagues draw from this assessment again follow those that economists have made for some time—federal policy should give greater weight to redistributive concerns, and less weight to developmental ones. State and local governments perceive themselves in competition with each other for jobs and upper-income households and are likely to emphasize developmental efforts and downplay redistributive programs unless there is federal funding for programs aimed at lower-income groups and strong and persistent federal political support for local redistributive program professionals.

There is much of value in this book for both scholars and students. Peterson, Rabe, and Wong are particularly convincing in their discussion of the political evolution of redistributive program management, showing in considerable detail how federal and local managers in health and education were able to overcome initial problems of high expectations and limited resources and develop effective programmatic and political relationships.

The discussion of the generic features of developmental programs is less convincing. There are numerous cases in the literature of considerable local controversy surrounding supposedly popular developmental programs, ranging from relocation disputes in urban renewal and highway programs to so-called not-in-my-backyard issues that frequently arise in connection with the location of roads, sewer plants, and airports; and there are equally numerous cases, including some in the Community Development Block Grant program, which is studied here, of developmental policy professionals controlling the operation of programs in the face of indifference or even active opposition from elected officials. As in the case of most other program typologies, what is developmental and what is redistributive is frequently more in the eye of the beholder than in any distinctive program characteristics.

More generally, there is less attention to the role of policy professionals in nominally developmental areas than might have been expected, particularly given the importance attached to their role in redistributive areas. Much of the limited success that Peterson, Rabe, and Wong note from attempts to attach redistributive features to developmental programs may stem from opposition from program policy professionals who see their function as building things in the straightest line at the lowest cost or simply see their professional prestige as jeopardized by having to take welfare recipients as clients. What is advantageous in one policy area may be the source of political tension and conflict in another.

JAMES W. FOSSETT
University of Illinois
Urbana

RIKER, WILLIAM H. *The Development of American Federalism*. Pp. xiii, 233. Boston: Kluwer Academic, 1987. No price.

With the exceptions of the introduction to the book and the commentary on each chapter, this book is composed of materials previously published in book or journal form or prepared for conferences. The latter materials are published for the first time in this book. Chapter 5 is based on the Ph.D. dissertation of a former student, William P. Alexander, who died before he had the opportunity to abridge the dissertation for publication as a journal article.

Chapters are devoted to the origin of the federal government, the invention of centralized federalism, Dutch and American federalism, disharmony in the federal government, the measurement of American federalism, the relation between structure and stability in federal governments, the United States Senate, the decline and rise of the militia, administrative centralization, presidential action in congressional nominations, and party organization. Although written at different times for different purposes, the chapters generally tie together reasonably well and are based upon the general theme of the continuity of American federalism.

The most interesting statements are found in the newly written introduction to the book, in which Riker describes his "ideological migration . . . from New Dealer in the fifties to liberal, anti-statist in the eighties." Initially, Riker viewed federalism in the United States as "an impediment to good government" but subsequently came to "regard it as a desirable, though still minor, restraint on the leviathan."

The major problem with the book is the lack of an epilogue updating the various chapters and tying them together. After reading this relatively short book, we know little about its author's changing views on American federalism beyond what he wrote in the new introduction to the book. Expansion of his views on the change in the moral meaning of federalism would have added greatly to the book's value.

JOSEPH F. ZIMMERMAN
State University of New York
Albany

ROBINSON, DONALD. *"To The Best of My Ability": The Presidency and the Constitution*. Pp. xvi, 318. New York: Norton, 1987. $22.50.

This is an interesting and valuable book, certainly no less so because its author is a protégé of James MacGregor Burns. In fact, Donald Robinson, who teaches at Smith College, lives equidistant between Northampton and Williamstown, Massachusetts, so as to be close to and have more ready access to his mentor. That being the case, one wonders if the mentor approves of his protégé's conclusion that the American presidency and Congress are so deadlocked that the constitutional system needs radical surgery—to the extent that (1) presidents should appoint members of Congress to their cabinet; (2) presidential and congressional elections should coincide every four years—with concurrent terms, except for Senate terms of eight years; (3) special presidential and congressional elections should take place in times of deadlock, dissolutions to be effected by Congress with or without concurrence of the president; and (4) a national council of 100 notables, serving for life, should issue calls for special elections and superintend such elections. Robinson's proposed constitutional system borders on parliamentary government except for two important differences: the concept of separation of powers is maintained; and the president and Congress would continue to have different constituencies.

Robinson's account begins with the development of the American presidency, moving rather conventionally from the original idea of the executive, through early experiences in state constitutions, "the Convention of 1787," and "the Chief Executive" from 1789 to the present. He then addresses various constitutional aspects of the presidency, such as elections, war powers, bureaucratic structures, law enforcement, and the management of the economy.

"*To The Best of My Ability*" is well grounded in classic writings as well as the important secondary sources. Based on such sources, the book would stand alone as a contribution to understanding the development of the presidency. The illustrations of recent "deadlock of democracy"—national parks policy, economic stabilization, antitrust prosecution, civil rights, and air pollution—are revealing and well narrated.

It is, of course, Robinson's conclusion that particularly attracts attention. At first blush, its prescriptions verge upon fantasy, as have those of many other students of the presidency over 200 years. But on second thought—given the recent numerous stalemated relations between the president and Congress noted earlier—what better way to break a deadlock? Need we be wedded to the constitutional system of the fathers? Perhaps, in the Constitution's bicentennial year, we celebrated too ardently the sacredness of the framer's constitutional arrangement. The case may be that the nation's ability to dissolve its government voluntarily in times of stalemate may one day result in involuntary dissolution. "It can happen here."

MARTIN L. FAUSOLD
State University of New York
Geneseo

SIMONTON, DEAN KEITH. *Why Presidents Succeed: A Political Psychology of Leadership*. Pp. xi, 292. New Haven, CT: Yale University Press, 1987. No price.

This book makes bold claims about predicting presidential success. Here is one such example: "We can predict presidential success with an acceptable and sometimes imposing precision." And if presidential success is defined in terms of gaining sufficient electoral votes to succeed in the general election, the claim sounds even bolder: "On the basis of a small collection of predictors, the victor in the electoral college can be forecast without mistake." Finally, the claim is made that "the incumbent's popularity in polls of the American people likewise can be accurately predicted using a handful of variables."

For the person who holds a worldview that allows for randomness, free choice, and serendipitous outcomes, these claims seem a bit arrogant. If outcomes of presidential elections, standings in political polls, and presidential ranking in history—all measures of success—can be so readily known in advance, then why do we continue to be surprised when a Jimmy Carter, a Ronald Reagan, or even a John F. Kennedy is elected president? And even if we granted that there were clear-cut criteria for presidential success, we might want to ask, Do these criteria name necessary or sufficient conditions? Let us suppose that a candidate for president satisfies a few of the criteria for presidential success. Among these are

—integrative complexity, or the capacity to make fine differentiations and com-

prehensive integrations of information;
— intelligence;
— height—taller individuals are more likely to be elected than shorter ones;
— academic or military career;
— previous political experience on Capitol Hill; and
— broad exposure to the American people.

Now even if we could identify a candidate who satisfied all of these criteria plus others that Simonton identifies as essential to electoral or historical success, would they be sufficient? Suppose Gary Hart satisfied all of these. But he was also a womanizer. Here the criteria are necessary but not sufficient predictors of electoral success. Suppose Joseph Biden satisfied all of the criteria. But he also had a penchant for plagiarism. Again, because such indicators are not in the equation, it would seem virtually impossible to predict what the candidate's fate would be. Or take Richard Nixon. If an event of Watergate's dimensions had occurred earlier or if leading Democrats could have uncovered more information sooner, would Nixon have been reelected?

One does not have to be a quibbler to raise such doubts. Commentators do not always agree on the kinds of things that are relevant to predicting presidential success. For example, there is a perfect correlation between the amount of money a candidate spent on an election and election success for the past six presidents. Does that mean the candidate who spends the most money to get elected will be elected? Could McGovern have been elected, no matter how much he spent? There is a high correlation between good presidential management of the economy and reelection for presidential incumbents. Does that mean a president will automatically be reelected if he or she manages the economy well or if it is in good shape? Would Johnson have been reelected no matter how well he managed the domestic economy, given the national abhorrence for the Vietnam war? These questions abound because variables that seem important become isolated and prediction seems possible, but then something unusual happens and forecasters attempt to account for the unexplained outcome by altering the prediction model. Perhaps it is unfair to judge this book in that light, but I found no convincing evidence here that the science of political prediction has reached the refined level Simonton claims it has.

STEPHEN W. WHITE

Auburn University
Alabama

THERNSTROM, ABIGAIL M. *Whose Votes Count? Affirmative Action and Minority Voting Rights.* Pp. xii, 316. Cambridge, MA: Harvard University Press, 1987. $25.00.

The revolutionary nature of the federal Voting Rights Act of 1965, a nationally suspensive law, is recognized fully by Abigail M. Thernstrom, who raises important questions about the act's transformation by judicial and administrative interpretations and its implications for the polity.

The act requires covered states and political subdivisions to submit electoral changes, including laws enacted by the state legislature and signed by the governor, to the U.S. attorney general or the U.S. District Court for the District of Columbia for approval before the changes can become effective.

Thernstrom directs harsh criticism at the Voting Rights Section of the Justice Department: "Administrative decisions have diverged from those of the Supreme Court.... The Justice Department's decisions appear to follow no principle," and the department "has been creating detours around the law, but generally to a common and clear end." In addition, Thernstrom writes, "The decisive work is often done by employees without legal training or, in some cases, even a college degree."

As Thernstrom correctly notes, the administration of the act is based upon the assumption that whites cannot represent blacks and that blacks and Hispanics should be guaranteed representation in accordance with their population in a given state or local government by means of manipulation of electoral systems.

The United States Supreme Court does not escape without criticism. In particular, the Court can be faulted for upholding what David I. Wells termed an "affirmative" racial gerrymander in the Hasidic Jews case. Furthermore, Thernstrom writes, "by failing to define democratic representation, the Court has neglected to identify inadequate electoral opportunity." Based in part upon signals from the Court, the Justice Department is seeking to replace the at-large electoral system favored by early municipal reformers to rout out corruption associated with the ward or single-member district system.

Although the book can be criticized for failing to examine in detail the impact of the Voting Rights Act in covered jurisdictions, such criticism is unfair, as an examination of this nature is beyond the scope of a single book focusing upon the politics of the act's enactment and interpretation.

While Thernstrom explains the original goals of the act's drafters, she fails to identify the reason Texas in 1965 purposely was excluded from the act's coverage, thereby necessitating the legislative invention of a second trigger—a voting device—to make the suspensive act apply to a given subnational unit.

It also would have been helpful if she had included a short section on the proportional representation electoral system as an alternative to the single-member district system to increase direct representation of blacks and language minorities on elected bodies. The proportional representation electoral system permits representation of minorities in direct proportion to their respective vote totals should they believe they can be represented only by members of their groups, yet it does not suffer from the disadvantages that led to the discrediting of the ward system early in the century.

In sum, the book can be read with profit by specialists in electoral politics and by the general public. Not everyone will agree with Thernstrom's analysis. The civil rights lobby, acknowledged by Thernstrom to be highly effective, no doubt will dispute her findings.

JOSEPH F. ZIMMERMAN
State University of New York
Albany

SOCIOLOGY

CLAVEL, PIERRE. *The Progressive City: Planning and Participation, 1969-1984.* Pp. xviii, 234. New Brunswick, NJ: Rutgers University Press, 1986. $28.00. Paperback, $10.00.

Pierre Clavel has written an important book. Looking at the progressive city governments that developed in Hartford, Cleveland, Berkeley, Santa Monica, and Burlington in the 1970s and 1980s, Clavel sees much more than these five cities. He raises fundamental questions about the relationship between planning and participation, the nature of progressive politics, and the goals and orientation of social science research. His search for answers centers on the need to develop new and productive forms of participation in planning, politics, and academic research.

What Clavel finds in these progressive cities is a strong linkage between planning and genuine participation. Professionals and community members genuinely learned from each other and interacted, in many instances, as equal partners working to change their city for the better. Active interaction and engagement encouraged creativity and indeed helped planners to function simultaneously as realistic generalists and effective problem solvers. These cities provide, for Clavel, concrete examples that planning need not be an "elite, bureaucratized activity." The experience he describes conforms much more to his noble vision of the profession as one that "long harbored a loosely structured ideal of moving directly among the people . . . and one that, in principle, was dedicated to visions and models that could be validated by catching the popular imagination."

Clavel emphasizes that progressive city

governments, no matter how well intentioned, are often unable to sustain a participatory approach to planning and societal change. The defeat of progressive candidates in both Cleveland and Hartford are attributed to a failure to maintain connections to neighborhood, grass-roots constituencies. It is to Clavel's credit that he sees community participation as much more than a necessity for political success. Identifying participation as the basis of progressive coalitions, he writes, "People, getting a taste of political participation, desired more. They did not change their attitudes or behavior directly in response to such factors as police violence, but rather they changed in response to the experience of political participation itself." The crucial issue, as Clavel recognizes, is how to encourage, nurture, and sustain appropriate and effective participation that promotes effective government and enhances societal well-being.

The issue of effective and appropriate participation also underlies Clavel's discussion of the nature of social science research. Criticizing what might be termed the paternalistic, expert model of university research, his work calls for and exemplifies an approach that involves learning from the community. Responding to William Foote Whyte's 1981 presidential address to the American Sociological Association, in which he challenged sociologists to discover, describe, and analyze "social inventions for solving human problems," Clavel conceptualizes his role as one of reporting on and interpreting a new form of social organization. Given that progressive governments could significantly benefit from the recording and circulation of information about their own work, reporting, for Clavel, serves not only scholarly but also practical purposes.

While Clavel is to be commended for his emphasis on learning from the community and his desire to contribute to the social movements he supports, he does not go far enough. Clavel defines his role and that of university researchers in general as the essentially passive and reflexive one of "constructive reporting." University researchers, in my judgment, can take a more active role and in so doing make potentially more significant contributions to social science and society. They can engage in what William Foote Whyte in *Learning from the Field* (1984) termed "participatory action research."

Participatory action research is likely to attract an increasing number of scholars in the years ahead. Clavel's description of planning in his five progressive cities has much to teach those interested in this promising approach. The participatory, interactive, collegial, problem-focused orientation that developed between planners and community members provides a most suggestive model for linking social science research with action and policy. As Clavel makes clear, that participatory orientation is part of a wider trend away from rigid hierarchy toward democratic participatory forms. Indeed, his work can be seen as a most important contribution to our general understanding of phenomena as seemingly diverse as quality circles in business, recent teacher-centered reforms in education, and the partnerships developing between universities, as well as other institutions, and their communities.

IRA HARKAVY

University of Pennsylvania
Philadelphia

ESTRICH, SUSAN. *Real Rape.* Pp. 160. Cambridge, MA: Harvard University Press, 1987. $15.95.

Susan Estrich has written a timely and important book. Her subject is simple rape—those rapes perpetrated by a lone man, known to the victim, who enacts his crime without the aid of a deadly weapon. Simple rapes are distinguished from those rapes that fit the stereotype of rape—rapes committed by an unknown, armed assailant who attacks a woman and leaves her with visible physical injuries. There is compelling evidence from social scientists that simple rapes constitute the vast majority of rapes and that they are responsible for serious and often long-term psychological effects. Legally, distinctions between the two types of rape revolve around aggravating factors such as the presence or absence of a weapon; both are equally criminal when committed through force or threat of force and without the woman's consent. As Estrich compellingly documents, however, in practice there is a significant and consistent distinction made between these two rapes. Rapes that conform to social expectations of "real rape" tend to be reported by victims, vigorously prosecuted, and harshly punished. Simple rapes, on the other hand, are often not reported to authorities, when they are reported they are disproportionately "unfounded" by police, are less likely to be accepted for prosecution, and, when successfully prosecuted, run a good chance of being overturned on appeal.

It is this difference between simple and aggravated rape, and the social implications of this distinction, that form the basis of this book. Using as her primary data the writings of appeals courts as they define the parameters of applied law, Estrich documents the systematic bias with which cases of simple rape are treated by the legal system. In general, Estrich argues that the legal view of rape victims, particularly victims of simple rape, is distinguished by its unrelenting and unfounded suspicion. She presents evidence to support her contention that the underlying sexism is so powerful as to render illogical and contradictory many legal rulings. For example, she compares the evaluation of rape charges and the standard of evidence with other crimes such as robbery and demonstrates a significant discrepancy. In rape, the burden of proof—for example, consent, resistance—is placed on the woman's behavior rather than on the defendant's.

She goes on to analyze some of the underlying assumptions, expectations, and projections of the men of the legal system as they apply the law to victims of rape. For example, she notes that the level of resistance required of a woman in a simple rape case completely ignores the usual size and strength disparity between men and women as well as the different socializations that prepare males and females to use their bodies in very different ways. Women are expected to be passive, physically timid, and sexually constrained and to protect their chastity even when circumstance and logic indicate that doing so may further endanger their lives. Yet they are also expected to fight like a marine when someone they trust tries to abuse them. "Their version of a reasonable person is one who does not scare easy, one who does not feel vulnerable, one who is not passive, one who fights back, not cries. The reasonable woman, it seems, is not a schoolboy 'sissy'; she is a real man."

In addition to analyzing the current application of rape laws, Estrich provides a thoughtful and critical discussion of the new rape reform effort. She cogently argues that many of these feminist-influenced reforms are backfiring because they obscure the "unique indignity" of rape. These reforms are failing, she suggests, because not recognizing difference where difference exists constitutes discrimination just as surely as constructing difference where there is none. Thus the laws need to take into account power and strength disparities between men and women just as they need to acknowledge women's rights to their own bodies.

Estrich argues that the legal system is a powerful transmitter of social values and expectations. Thus the legal system may be used to perpetuate and legitimize fallacious and sexist views of women, or it may reflect progressive trends in the culture and act as a powerful agent for social change. Her closing suggestion is that the laws should reflect and hence promote changing expectations of female autonomy and human rights. Specifically, she suggests that the practice of the law should enforce a woman's right to say no to unwanted sex. In doing so, she hopes that "the legitimatizing power of the law [may be used to] reinforce what is best, not what is worst in our changing sexual mores."

Estrich's book is very well written. It is succinct, direct, and readable. It is perhaps something of an accomplishment in its own right to write a book about the application of the law that is accessible to the nonlawyer. It is perhaps even more notable to write for a broad audience about issues of gender and social structure. This book is appropriate for a general audience, and a must for people interested in a sensitive analysis of the relationship between social institutions and human behavior.

LESLIE LEBOWITZ
Duke University
Durham
North Carolina

LANDRY, BART. *The New Black Middle Class*. Pp. xi, 250. Berkeley: University of California Press, 1987. $22.50.

The New Black Middle Class is a discursive report on a 1976 survey of 556 black two-parent families, often employing comparisons to an identical survey of 600 white families. Landry employs occupation-based criteria, rather than education-based or income-oriented ones, for determining the middle class, and hence his subjects include clerical and sales workers while excluding the sometimes better-paid blue-collar trade workers.

If Landry's initial criteria for class raise questions and potential problems, the book itself does little to alleviate those concerns or avoid other shortcomings. The essence of Landry's argument boils down to some four central points. First, "the black middle class ... is a kind of bellwether of black progress." Second, the black middle class experienced phenomenal growth during the decade of the 1960s, increasing from 13 percent of all black workers in 1960 to 27 percent in 1970. Thereafter, that growth substantially slowed; in 1976, 31 percent of black workers were middle-class, compared to 53 percent of whites, the latter being nine points higher than the 1960 figure of 44 percent.

Third, Landry says, that impressive growth in the 1960s resulted from the simultaneous presence of a strong national economy and the civil rights movement. More precisely, it was federal antidiscrimination statutes, particularly Title VII of the 1964 Civil Rights Act, that supplied the crucial momentum for those gains. "It was these new laws mandating equal employment opportunities that had the most far-reaching consequences for black people as a whole and that directly contributed to the growth of the new black middle class."

Fourth, in a very poorly supported effort to critique the work of William J. Wilson, Landry contends that the slower post-1970 and especially post-1980 growth of the black middle class is explainable more by reference to racial discrimination in employment than by nondiscriminatory economic factors and changes. Indeed, such discrimination affects all classes of black workers and not simply the middle class. "The income gap between black males and white males is *primarily* the result of discrimination," and "occupational discrimination continues to be at least as important as economic change in maintaining a large black underclass." Hence, Landry concludes, "economic equality for blacks, even for the middle class, now seems a long way off."

These assertions are all arguably true, but Landry unfortunately provides no data or analysis that persuades his reader that these are conclusions drawn from his study rather than simply from personal beliefs. In fact, at times, some of Landry's most interesting statistical statements appear to contravene his more general argument, and the contradiction is not explained, as in the following: "Between 1973 and 1982 ... the black middle class grew by 51.4 percent compared to 28.7 percent for the white middle class," and "during the 1980-1982 recession, the black middle class grew by 3.4 percent and the white middle class by 2.9 percent."

In short, I found *The New Black Middle Class* to be a fundamentally unsatisfying and disappointing volume.

DAVID J. GARROW

City College of New York

City University Graduate Center
New York

TRAUTMANN, THOMAS R. *Lewis Henry Morgan and the Invention of Kinship*.

Pp. xiv, 290. Berkeley: University of California Press, 1987. $30.00.

The letters "F.M. & B.M.S." standing for "For more and better Morgan Scholarships" were originally used to conclude a letter from Professor Tooker, a Morgan scholar, to Thomas R. Trautmann. This motto probably best describes the intentions that led to the writing of the book under review. But it is not so much Lewis Henry Morgan (1818-81) as a person that forms the focus of this volume but one of his books, the influential and controversial *System of Consanguinity and Affinity of the Human Family* (1871). As Trautmann puts it, he intended to write a biography of Henry Morgan's *System* and by all accounts he succeeded splendidly.

There is little doubt that with Morgan, Tylor, and Spencer, anthropology came of age. That Morgan, after receiving initial acclaim, slipped into disfavor after the turn of the century must be attributed to the philosophical élan, not altogether unknown in the second half of the nineteenth century, to fill the gaps of anthropological knowledge through grandiose conjectures. Morgan's failings in this regard have overshadowed a number of contributions that had an enormous effect on the development of anthropological thought and practice over the last century. Among these is the invention of kinship, or, better said, the invention of a classificatory kinship system as opposed to the natural or descriptive system. The first reduces consanguinity to great classes by a series of arbitrary generalizations and applies the same term to all members of the same class, for example, "my father's brother is my father." The second, descriptive or natural, system describes consanguinity through a combination of primary terms, such as "my father's brother." Calling these distinctions "a new instrument for ethnology," Morgan understood them to be analytical tools capable of tracing patterns of descent to their ancestral roots. But Trautmann does more than retell what is commonly known about Morgan. He elucidates Morgan's philosophical background, whose principles and axioms determined the direction that his scientific inquiries took. His theory of mind, for example, emphasizes the tutelage of nature to arrive at true knowledge. The activities of the mind, on the other hand, are likely to be arbitrary, artificial, and false. Marx and Engels took a keen interest in these formulations, coming as they did from a "Yankee Republican." "Morgan's ethnology, concentrating on the growth of knowledge, not so much through contemplation and reflection as through interaction with external nature in the production of subsistence, are in deep harmony with that philosophy." So are a number of theoretical directions in anthropology today.

Thomas R. Trautmann is meticulous in his research and has produced a well-organized exposition of one of the early giants in anthropology, whose fate deserves better than the half-forgotten label attached to him at the present.

KARL A. PETER
Simon Fraser University
Burnaby
British Columbia
Canada

WHITTAKER, ELVI. *The Mainland Haole: The White Experience in Hawaii.* Pp. xxxiv, 233. New York: Columbia University Press, 1986. $25.00.

The theme of the 1986 Hawai'i Sociological Association featured ethnic research on the mainland and Hawaii. The closing session of the meeting was inspired by Whittaker's volume and the controversy it created among ethnic scholars in Hawaii. Simply stated, Whittaker suggests that ethnicity is a social construction and that, in Hawaii, Haoles—recent white immigrants from North America—constitute an ethnic group.

In recent years, much attention has been given to the notion that Haole has become a stigmatized, undesirable group status in the islands. Viewed as outsiders, the group is blamed for past and present cultural, economic, political, and social woes. Moreover, the group is identified as the primary destructive force behind the decline of Hawaiian cultural values and practices. Nowhere is this view more prevalent than among Haoles themselves, although individually most claim full exemption from any and all historical responsibility. While many Haoles endorse this reconstructed view of history, it is true that many local groups do as well.

Whittaker's book purports to present the findings of an interpretive and humanistic exploration into the social construction of the Haole ethnic status. Replete with discussions of both positive and negative aspects, an investigation of interactive patterns, and social consequences, the volume is constructed around a preface, prologue, and five thematic chapters. Briefly, the preface and prologue are a series of disclaimers and qualifiers enmeshed in an elementary discussion of the metatheoretical issues involved in producing human science knowledge or theory. Chapter 1, "Dialectics of Enterprise, Hope, and History," claims to illustrate the universalized form of "Western consciousness" and its impact upon Hawaiian culture. Like diseases, this consciousness is traced historically through a series of white hosts both before and after Cook's arrival in 1778. Chapter 2, "Discovering the Haole," is a mixture of reflections on methodology and epistemology and on Whittaker's own feelings concerning doing research on Haoles in Hawaii. Chapter 3, "The Migration Story," explores the who and why of Haole migration to the islands. Chapter 4, "Nature as Mediated Metaphor," seeks to present the universalized Haole view of place or space. Finally, chapter 5, "Rituals of Inequality: Ethnicity and the Haole," begins the systematic presentation of the actual ethnographic data and offers some important insights into the real lifeworlds of Haoles. This is without question the best chapter in the volume for it takes up the meaty issues of how Haoles define, respond to, cope with, and survive in this allegedly hostile, or at least different or indifferent, social environment thought to be paradise prior to migration.

Whittaker's book raises more questions than it answers. In this instance, the Socratic and pedagogic function of such a tactic is neither fruitfully nor skillfully employed. For instance, the reader is never given to understand why the Haole status of today emerged when and in the form it did. There is no comparative data or analysis offered to help clarify this matter, and a systematic discussion of institutionalized power in Hawaii is completely lacking. While place is discussed, there is no phenomenological analysis of time as it affects the Haole migration experience. None of the discussions are grounded in a concern for what is happening in the larger Pacific Basin or world system contexts. There is never a full disclosure as to whether these Haole perceptions coincide with other groups' perceptions and experiences or with what can be gleaned from the historical record itself. Finally, the reader is never given to understand whether what exists today is the product of a gradual evolution in the social ecology of Hawaii or the result of some sudden internal or external crisis-induced structural change. These omissions are inexcusable especially for an interpretive or phenomenological anthropologist who claims that history is a central key for unlocking the significance of socially constructed statuses. If Whittaker had really taken advantage of the available data, she would most assuredly have known that Hawaii is in a period of social and political transition, a historical era wherein no ethnic group enjoys total institutionalized hegemony and the situation might best be described as ethnic anomie. Unfortunately, the reader is never informed of the actual historical realities of Hawaii and is led to believe that somehow what Whittaker says about Haoles is true independent of all ongoing historical dynamics.

Whittaker's discussion of Western consciousness, Haole views of place, the various migration histories, and Haole tendencies to universalize their life-worlds never really seems to jibe with her ethnographic data. It is almost as though there were two books here, one an oversimplified discussion of Western cultural values and the other on the title theme.

On balance, this is a disappointing book. Whittaker has written on an important topic, with data from an ideal social laboratory, and has referenced a series of important—albeit controversial—ideas without really taking advantage of any of them. She has basically written a book about doing research and not a book on the important research she actually did. Ironically, "Haole" originally meant "without breath" and was intended to describe the standoffishness of white foreigners as they interacted with native Hawaiians. In this historical respect, Whittaker has indeed written a very Haole book.

JEFFREY L. CRANE
University of Hawaii
Hilo

ECONOMICS

AMY, DOUGLAS J. *The Politics of Environmental Mediation.* Pp. x, 255. New York: Columbia University Press, 1987. $30.00.

BOSSO, CHRISTOPHER J. *Pesticides and Politics: The Life Cycle of a Public Issue.* Pp. xvi, 294. Pittsburgh, PA: University of Pittsburgh Press, 1987. $29.95.

These books provide good insight into academic research concerning environmental politics. A strong conceptual framework is combined with substantial documentation. Both Amy and Bosso are political scientists whose natural orientation is understanding the origins, processes, and outcomes of political change. Thus Bosso states, "I am less interested in judging the virtues of federal pesticides regulation than in how politics affects policy, and vice versa. . . . My main goal is to understand . . . political change." Amy is concerned with "the issues of power, equality, and democracy that are necessarily involved in any policymaking process." Environmental politics is simply a specific power arena.

Each book is in the nature of a careful and interesting case study. Bosso examines the evolution of federal pesticides regulation from the 1947 Federal Insecticide, Fungicide, and Rodenticide Act (FIFRA), passed to protect farmers. This work is an artfully crafted longitudinal study that contrasts a systematic political analysis with the evolution of federal regulation. Amy criticizes informal mediation—site-specific dispute resolutions, regulatory negotiations, policy dialogues—as a means for achieving compromise on environmental controversies. Bosso's and Amy's interests in the environmental politics arena do not appear to be value free. I infer that each author views the American governance system as being overly biased toward development interests.

This position is very clear in the instance of Amy. He concludes that "environmental mediation should occupy a relatively minor role in environmental politics" relative to litigation and legislation—administration and binding arbitration are the other possible options—because mediation is a "subtle but powerful form of political control." Development interests—usually large corporations—will be better prepared for the informal negotiation process, which will thus be biased in their favor: "Mediation can be no more fair than the larger political context within which it takes place," writes Amy. Mediation proponents have generally cited its greater speed, lower cost, and capacity for generating compromise. "The most important political issue surrounding environmental mediation is not how fast or cheap it is, but how fair and just it is." The issue is not better communication between opposing interests: "Politics is not simply about communication; it is also about power struggles." Mediation would work best under the relatively rare conditions of relative balance of power and thus political impasse. This book argues for confrontational approaches as devices to enhance the power base of the relatively weak, who are largely those opposed to or affected by development.

Chapter 1 of *Pesticides and Politics* is an excellent summary of the debate in the literature on U.S. governance. It systematically compares the elitist, pluralist, subgovernment—"iron triangles"—and issue-network interpretations. Bosso examines longitudinally the structure of policy bias and its impacts in the environmental politics area. The participation revolution has led to more competition over time between policy claimants. "If the pesticides case shows us anything, it is the emergence of an interest-group population dramatically different from that studied by pluralists in the 1950s." Bosso also concludes that environmental politics is one example of what he terms "intractable"—and therefore continuing—policy problems. The problems of nuclear power, toxic waste disposal, and pesticides regulation are so morally and technologically complex and so central to modern society that endless controversy must be envisioned.

While both books can be readily recommended to specialists in environmental politics, and while Bosso's summary of the democratic governance literature is excellent, I doubt that they constitute major contributions to the democratic theory literature. Rather, they are careful but not innovative applications of that literature to the environmental arena.

DUANE WINDSOR
Rice University
Houston
Texas

ARNDT, A. W. *Economic Development: The History of an Idea.* Pp. viii, 217. Chicago: University of Chicago Press, 1987. $20.95.

Many are the areas of public concern that have faced the post-1945 world. Few, however, even approximate the concern for economic development, especially in what we have come to call the Third World.

Economic Development, written by A. W. Arndt, professor emeritus of economics at the Australian National University and author of the important work *The Rise and Fall of Economic Growth*, attempts to trace the history of the idea of economic development as a policy objective in less developed countries. It succeeds very well in that attempt, comprehending the diverse as well as complex dimensions the idea has assumed in its evolution.

Arndt's approach may be captured under four headings: the source of the idea he investigated; its emphasis, from the standpoint of policy; its nature; and the social context within which it has evolved. We will look at each in turn, seeking as we do so to incorporate some substantive aspects of the work.

As respects the sources, Arndt focused on two areas, as he canvassed information on the idea of development—academic writings and public debates. He covered the range from Adam Smith to Ayatollah Khomeini. With respect to the emphasis assigned to the views of development, Arndt bypassed means and concentrated on ends, detailing shifts in the latter from capital formation to growth and equity.

The nature of the ideas with which Arndt dealt is one that is principally economic but one that also embraces thinking in other fields that bear on the idea of development. Thus the volume houses ideas ranging from those emanating from the claims of liberation theology to those based on the ethical teachings of Gandhi. Finally, the evolution of and the varieties in the idea of development are firmly planted within their sociopolitical contexts, thus enabling one to appreciate more keenly certain political differences even among states sharing a common ideology—the USSR and China, for example.

The work makes an important contribution to the literature on economic development, especially as it incorporates ideas on a theme that informs our concern for social justice, individual and social freedom, identity, and community. It is written in clear, nontechnical language; the sources from which it draws are representative; and although some will find portions of it familiar because they were published elsewhere, as a whole it should be instructive to all who are interested in international affairs.

WINSTON E. LANGLEY
University of Massachusetts
Boston

GALENSON, DAVID W. *Traders, Planters, and Slaves: Market Behavior in Early English America*. Pp. xiv, 230. New York: Cambridge University Press, 1986. $34.50.

This appealing example of the application of statistical technique to a topic in economic history deals with the transit traffic in slaves to the early British West Indies. David Galenson first reminds readers of the central importance of the West Indies to British transatlantic trade at the start of the eighteenth century. In the year 1700, Barbados had the greatest population of all the British American colonies, and there were nearly twice as many slaves on that island as on all the British mainland. In the years round about 1700 (1697-1705), the mean annual value of imports from the West Indies was more than double the value of imports from the North American mainland colonies, with Barbados alone nearly matching the total for the entire mainland.

This prosperity was based on the huge potential for sugar production, a potential that would not have been realized with a white labor force. The severe shortage of white indentured labor was due to high transport costs and poor living conditions for workers in the West Indies, together with the hard and unpleasant tasks that marked the cultivation of sugar. A steady supply of slaves at reasonable prices was essential.

Over a period of some 50 years following its establishment in 1672, the Royal African Company (RAC) was one of the largest participants in the slave trade, attempting to exploit its leaky monopoly status that expired in 1712. Galenson has chosen a number of topics concerning the RAC's trade that can be illumined with statistical analysis. His is a very different book from Kenneth G. Davies's well-known study, *The Royal African Company*, published in 1957, as that study was comprehensive.

Galenson has selected for study a number of areas where the evidence—taken from the 24 volumes of accounts that survive in the company's records—indicates careful attention by the RAC to rational economic calculation within a large, competitive, and complex market, contrary to claims of imprudent mismanagement that date at least as far back as Adam Smith's *Wealth of Nations*.

Owners of ships hired by the company, for example, were required to take a portion of their earnings—a minimum of one-quarter and often as much as two-thirds—in slaves. Shipowners thus disposed of nearly 20 percent of all RAC slaves brought to the West Indies, easing the company's marketing tasks. Payment for hired ships was a flat sum per slave delivered alive, with no payment for deaths in transit, an effective way to reduce losses from poor shipboard conditions on the slavers.

Evidence is presented that shipments were seasonal, meant to avoid the unhealthy rainy season in West Africa and the hurricane season in the West Indies. In wartime, there was a shift to heavier vessels that could carry a larger armament, and to smaller ones that could evade pursuit. By boosting the costs of transport, war also led to a reduction in the share of children in slave cargoes. Higher prices for slaves led, on the contrary, to a larger share of children in cargoes, while lower prices accomplished the reverse. The correspondence of RAC officials indicates that these consequences were not accidental: the officials were well aware of these developments.

One finding by Galenson is that slave prices consistently declined while the RAC sales were in progress. He explains this by tying prices to quality, arguing that the highest-quality slaves were the first to go to the block. The so-called seasoning of unhealthy slaves accounts for some of this, with the seller unwilling to make public early in the sales that contagious disease was afoot. The main reason identified by Galenson, however, is that wealthy planters with a high premium on their time took the best-quality slaves early in the sales, while the poorer planters took the leavings. He suggests, rightly, that the ordering of purchase and sale by quality occurs in many other markets as well, a phenomenon that calls for more analysis.

In another research area, Galenson investigates the "persistence" of West Indian planters at the RAC's sales, using the results to infer conclusions concerning death rates and rates of out-migration. The result is somewhat surprising: over some time periods, the "persistence" data indicate greater stability in the population than suspected by narrative historians.

In the end, the company failed neither for lack of effort nor because of bungling mismanagement, but for intrinsic and understandable economic reasons. These reasons were (1) the wretched communications that connected the company's agents in London, West Africa, and the West Indies; (2) the necessity of maintaining, under its charter, the very expensive forts and factories on the West African coast, while at the same time, the British government failed to police the company's monopoly against interlopers; and finally (3) the chronic shortage of currency in the West Indies, meaning that the company was forced to extend a great deal of credit there, encountering at the same time serious problems in recovering bad debts. All three of these together proved to be a fatal burden.

Traders, Planters, and Slaves is an outstanding application of detailed statistical analysis to a set of problems in economic history that will be well received by students of the slave trade.

JAN HOGENDORN
Colby College
Waterville
Maine

HIBBS, DOUGLAS A. *The American Political Economy: Macroeconomic and Electoral Politics in the United States.* Pp. xiii, 404. Cambridge, MA: Harvard University Press, 1987. $35.00.

In *The American Political Economy*, political scientist Douglas Hibbs constructs a useful framework for the dynamic relationship between electoral demand for and a partisan administration's supply of economic policies. This conceptual scheme is substantiated by both contextual details and vigorous empirical testing. One finds in Hibbs's presentation a systematic survey of the nation's economic performance—for example, real growth rate—in the postwar decades, an exhaustive review of the literature in both economics and political science, an elegant use of modeling techniques in the empirical testing of key propositions, and a comprehensive presentation of useful information on economic policy tools and the electoral process.

Above all, Hibbs restores a once dominant theme in the study of American politics: that political parties perform a central policy role. At a time when much of the literature suggests a decline in the relevance of political parties, Hibbs identifies partisan cleavage as the crucial variable in understanding macroeconomic policymaking. On the demand side, while the Democratic constituency, whose core members include blue-collar workers and lower-income classes, is more concerned with unemployment, the Republican core electorate—that is, white-collar workers and higher-income classes—seems more sensitive to inflation. These contrasting class-based preferences have shaped the supply side, or the partisan administration's policy objectives. As Hibbs observes, "Democratic presidential administrations typically pursue more expansive policies yielding lower unemployment and higher real output and growth ... than do Republican administrations."

To be sure, presidents are subject to numerous constraints in making economic policy. Otherwise, all presidents would succeed in timing their best economic performance in ways that would maximize their electoral gains, as the model of the political business cycle would have assumed. Both the Nixon 1971-72 recovery and the Reagan 1983-84 recovery were clear examples of how the political business cycle should work. More often than not, however, such a cycle seems to have gone in the wrong direction, as suggested in the 1979-80 ill-timed economic recession under Carter. These inconsistencies led Hibbs to conclude, "Election-oriented economic policy and output cycles have not been a pronounced feature of the American political economy." Instead, politically preferred economic performance can be undermined by class-based demands from the party's core constituency, domestic institutional constraints, the structure of economic relations, and the international economy in an increasingly interdependent world market—as occurred, for example, with the oil supply level determined by the Organization of Petroleum Exporting Countries. Hibbs concludes that, because these constraints are not likely to be altered by the president, "domestic politics and policy strongly influence but do not completely shape macroeconomic outcomes."

Much of the explanatory power of Hibbs's framework depends on the extent to which partisan class cleavages remain policy relevant and electorally important. In this regard, contemporary electoral behavior seems to suggest a gradual weakening in what Hibbs labels the Democratic core constituency. Since the 1960s, support among southern whites for Democratic presidential candidates has persistently declined pretty much regardless of economic circumstances. Even Jimmy Carter, a southern candidate, only received 46 percent of the southern white votes in 1976, as opposed to Ford's 52 percent. Given these political changes in the South, the relations between electoral outcome and macroeconomic performance seem somewhat less certain. Perhaps Hibbs should have elaborated on the implications of his framework of party realignment theories and the median-voter literature in his somewhat brief section on Reagan's legacy.

This study not only confirms some of our conventional thinking on economic policy-making but also offers thought-provoking analysis on this key policy area in the context of democratic process and structural constraints. In the aftermath of Black Monday on Wall Street in October 1987, this book offers timely insights into the role of government in the market economy.

KENNETH K. WONG

University of Oregon
Eugene

University of Chicago
Illinois

JENKINS, RHYS. *Transnational Corporations and the Latin American Automobile Industry*. Pp. xiv, 270. Pittsburgh, PA: University of Pittsburgh Press, 1987. $32.95.

In certain Latin American countries, governmental strategies favoring the creation of an automobile industry became a top priority in the late 1950s and 1960s. As a consequence, the industry came to be dominated by the major multinational enterprises that had previously served these markets through exports. They "produced high-cost vehicles in small, inefficient plants, whose output went to satisfy the demand of a narrow upper-income group." The first part of Jenkins's volume—some two-thirds of the book—carries the story to the mid-1970s. At that time, government strategies changed and, with balance-of-payments concerns, came to emphasize the promotion of exports rather than import substitution; this is the subject of the second part of the book. Jenkins believes that the new state policies, except possibly in Brazil, were not successful in creating low-cost production. Nonetheless, as this book shows, from 1975 on, Brazil has had a positive trade balance in the motor industry; in 1981, fully 27.3 percent of its vehicles were exported.

This study offers a fine introduction to the rise of the Latin American automobile industry, placed in the context of the world automobile industry of the last three decades. It has many valuable explanatory tables, albeit basic summary ones are absent; moreover, many tables are headed "vehicles," and the reader is not sure if they include cars and trucks, or only cars. No tables show, country by country, year by year, the course of automobile output, or cars per capita. It is only with a calculator and reading between the lines that the extraordinary relative importance of the Brazilian industry emerges. Jenkins writes of Mexico and Brazil's significance, yet Brazil produced in 1983 three times the vehicles made in Mexico. In 1983, Volkswagen in Brazil made more vehicles than General Motors or Ford did in all of Latin America. Jenkins does not fully explain why Brazil surged so far ahead of Mexico and particularly Argentina.

No tables show year by year the process of concentration in the motor industry, which comes out very vividly in the text. By 1983, the four largest manufacturers in Latin America—Volkswagen, General Motors, Ford, and Fiat—produced 82 percent of the vehicles made in the region. In 1980, Latin American output peaked at 2.2 million vehicles; in 1983, with economic recession, the figure was down to 1.5 million.

The automobile industry consists of what Jenkins calls the terminals—the producers of vehicles—tire companies, and the parts and components enterprises. Jenkins's discussion of the complex relationships between the terminals and the makers of inputs is masterful. My criticisms notwithstanding, all those interested in the world automobile industry or in the economic development of Latin America will want to read this book.

MIRA WILKINS

Florida International University
Miami

MARSHALL, RAY. *Unheard Voices: Labor and Economic Policy in a Competitive World.* Pp. xi, 339. New York: Basic Books, 1987. $19.95.

FLANAGAN, ROBERT J. *Labor Relations and the Litigation Explosion.* Pp. x, 122. Washington, DC: Brookings Institution, 1987. $26.95. Paperbound, $9.95.

Each of these policy-oriented books is a contribution to the theme that the regulatory institutions that govern our economy are no longer appropriate in their current form, having failed to keep pace with economic change. Marshall's book argues for increased participation of workers, the "unheard voices" to which the title refers, at all levels of economic organization as an economic imperative, if the United States is not to lose out in a world of increased global competitiveness.

This book is an argument for an industrial policy in the United States that Marshall is quick to distinguish from a planned economy, because an industrial policy would still be based upon and directed by the interplay of market forces. Still, it is Marshall's advocacy of a tripartite industrial policy that makes his book merit our attention. Few would doubt the benefits of increasing worker participation at the firm level, but Marshall's argument extends the merits of worker participation to the level of economic planning. He contends that economic policies such as those of the current administration that do not take into account the views of labor are ultimately damaging to our nation's economic health, because they are not based on the fairness and social justice needed to survive economic crises.

In Marshall's analysis, the threat to the U.S. standard of living is increasingly external, as a result of the increased competitiveness of foreign industry. Quite naturally, then, Marshall looks to the impact of foreign institutional arrangements on their industrial competitiveness, particularly in countries that successfully responded to external threats. It is looking abroad that gives Marshall most of his support for industrial policy. Japan, Germany, and Austria arose from wartime destruction because of the motive power of social consensus in rebuilding competitiveness.

Marshall's book is wide ranging and persuasive, and his advocacy of a tripartite industrial policy is tempered with the mindfulness that many readers may find such an institutional arrangement to be unknown or undemocratic in the United States. Marshall takes account of, if not fully rebuts, the criticisms of those who would argue that economic interest groups, particularly labor and management, are neither sufficiently well organized nor representative and thus that industrial policy would be unworkable and undemocratic. But it is the general weight of Marshall's argument for industrial policy, which is impressive, that must swamp specific objections to its institutionalization if the reader is to be persuaded.

His argument extends to discussion of problems arising from traditional practices at lower levels of economic organization. Here he describes the dysfunctional consequences of institutional arrangements that favor short-run decision making by managements. Traditionally hierarchical and authoritarian management styles and adversarial labor relations also impede economic growth because each either ignores or frustrates the contribution that the nation's work force can make toward economic renewal. Marshall gives persuasive evidence of the positive changes resulting from increasing industrial democracy, or worker participation, and economic democracy, or worker ownership. While change in this direction is encouraging, its pace must increase, which will only come about if participative structures receive institutional support. Unions will support increased labor-management participation if their own status is not eroded thereby, whence Marshall's recommendation to stiffen penalties under the National Labor Relations Act (NLRA).

The latter policy prescription provides the link to Flanagan's study, although Flanagan's analysis does not support the conclusions of those such as Marshall, who attribute the decline in unionization to the increase in illegal and legal challenges by managements of union representation. According to Flanagan, moreover, a change in labor policy that would increase the cost of violating the NLRA would probably result in an increase in management's positive labor relations strategies designed to increase worker satisfaction so workers will remain nonunion.

Flanagan concludes that the increase in regulatory litigation under the NLRA cannot be explained by changes in legal doctrine under the act or by an increase in the volume of labor relations activity subject to regulation, whether by the enlargement over time of the jurisdiction of the National Labor Relations Board (NLRB) or by regional and industrial shifts in labor relations activity. Rather, much of the increase in litigation—which has doubled in every decade since the 1950s—appears to be a behavioral response by labor and management to changing incentives to comply with the NLRA. Specifically, Flanagan finds that the growth of the cost of union relative to nonunion labor over the 1970s encouraged managements to violate the act and unions to file unfair labor practice charges, as managements had more to lose by complying, and unions, more to gain by filing under the act.

Although he discusses the range of policy options put forth in recent years, Flanagan favors partial deregulation as the best means of preserving NLRA rights and decreasing litigation. Delays in adjudication encourage increased litigation, because delays usually result in management victories. To break this cycle, Flanagan recommends research into NLRB rules that appear to have no relation to labor relations outcomes and the abolition of those that do not. Of these, Flanagan finds that the NLRB's "laboratory conditions" standard in the conduct of representation elections is irrelevant as voting behavior does not appear to be affected by illegal tactics; and that litigation arising under the "duty to bargain" is also unnecessary as it can be shown that the balance of power between labor and management, and not NLRB or court rulings, determines the outcome of bargaining.

Each book is a major contribution by a prominent author to central problems in our economic institutions. Flanagan's empirical work does not interfere with his book's readability, but both his prose and his argument are those of an economics professor, which may limit his readership, if not the book's influence. Marshall's work will appeal to a broader audience and will rightfully stimulate policy discussion especially as the Reagan administration draws to a close.

DUNCAN COLIN CAMPBELL
University of Pennsylvania
Philadelphia

PAUL, ELLEN FRANKEL. *Property Rights and Eminent Domain*. Pp. 276. New Brunswick, NJ: Transaction Books, 1987. $24.95.

In 1928, a man in Virginia was ordered to destroy valuable ornamental cedar trees because they harbored a plant disease that threatened a nearby apple orchard. No compensation was paid because the order was an exercise of local government to regulate—the police power—rather than the power of eminent domain. Ellen Frankel Paul has written a book about issues raised by this and numerous other cases involving government acts that diminish a property owner's rights.

Why one person's trees have to be sacrificed to protect another person's trees is a problem that would have troubled Solomon. Apparently King Solomon got off easy. He is not mentioned in this book. Many other familiar names, from Aristotle to Jay Forrester, play roles in Paul's treatise on the concept of property in relation to government. Aquinas, Locke, Bentham, Madison, Hume, Montesquieu, Marx, Rockefeller, and Nixon all have cameo parts. A good number of justices do, too. Governments of all stripes always seem to want to take or curtail the use of real estate that individuals think they own. "They can't do this to me!" the owners exclaim. Paul agrees.

The U.S. Constitution does not explicitly give government at any level the right to take or regulate private property; it does limit that right and so seems to presume that government will in fact do these things. And while two amendments speak of compensation for property that is "taken," the language leaves unclear whether or in what circumstances a land-use regulation—such as zoning or the order to cut down trees—is a compensable taking. This ambiguity has not deterred government, primarily at the local level, from imposing all manner of burdens on property owners for the benefit of the public. For example, ocean-front home owners in California have been required to allow public access to the Pacific.

Thus numerous land regulations have come before courts, giving rise to a tortuous trail of reasoning about how much government can do and why. While disappointing in other respects, Paul's book certainly conveys the flavor of a daunting legal conundrum.

For the most part, she claims, courts have deferred to the legislative branch when it comes to deciding what is required for the public good. If the city council says a person cannot operate a stone quarry, that is their privilege. They do not have to buy the person out. And right there is where Paul—and very likely many other people in our society—decide that enough is too much. Her way of putting it may be designed to shock: she wants the courts in this country to "stand as the final bulwark against government's propensity to seek the public good"—a fascinating point of view for a political science professor to take.

She admires the concept of property as a natural right, an old concept in fashion among our founding fathers when they disputed King George's dominion. Centuries before, King John's arbitrary taking had aroused his nobles; the Magna Carta gave the nobles back their domains. She attempts to clarify and, if you will, legitimize the natural right to property. Everyone needs property as a means of livelihood or at least they used to long ago. So a king or whoever should not take it away. Once that is settled, one property owner will not try to take away some other fellow's property because that would lead to no one's having anything.

At this point I looked in the index for "Native Americans"; they are not there, nor is anything about Bessarabia, Palestine, or Japanese Americans. How particular persons obtain their property in the first place is too delicate a matter for this book. All we need, says Paul, is a statute of limitations; whatever you took so far, you keep. King John would have loved that.

Having secured one's property in Paul's new world, no government could take it or tell one what to do or not do with it: not for road building, not on account of noxious externalities, not for anything. That is to say, no legislative body could encroach on a property owner's domain. But supposing a neighboring property owner does not like what his fellow owner's diseased trees are doing to his healthy trees? Sue! Paul wants courts to do after the fact what legislative bodies try to prevent.

To exercise a little critical prerogative, that just does not wash. Who would tolerate toxic wastes next door because he has the opportunity to sue after his children die? Will one driver live to sue the other driver who decides to take over the first driver's side of the road?

The concept of natural rights is no help in practical affairs. Both of my daughters claim a natural right to the bathroom at 7:30 a.m., the teenager because she was there—and here on earth—first, her little sister because she has to brush her teeth. Mom and Dad have to handle this with no help at all from John Locke. Neither Mother Nature nor James Madison has provided sufficient rules for day-to-day sharing of our environment. We have to improvise.

It is certainly very hard for anyone, judge or legislator or president, to be truly sure or really fair about what is in the public interest. That is not a reason for abolishing representative government.

WALLACE F. SMITH
University of California
Berkeley

RIDDELL, ROGER C. *Foreign Aid Reconsidered*. Pp. x, 309. Baltimore, MD: Johns Hopkins University Press, 1987. $35.00.

In his "long book both to read and to write" Riddle, a research officer at the Overseas Development Institute in London, provides unique, exhaustive, and often detailed analyses of the broad range of professional views on the theory and practice of foreign official aid to developing lands over the past three decades or so. He is concerned with the disciplines of economics and philosophy as well as with other social sciences and humanities as represented in over 500 source documents referenced in the volume and as addressed to goals of poverty alleviation and economic progress in the world's poor lands during the period.

The sources are individual monographs, including some that are considered accepted lore on the development process, as well as reports from major and active foreign aid agencies. They frequently relate theory to specific developing countries; most experience is reported by agency comparisons and explanations of very many individual aid project performance records. These sources embody approaches that are scientific and factual—at least quantitative—as well as those inspired by a wide range of ideological preconceptions and principles often garnished with presumably objective facts and logic.

On the whole, the book reveals an author whose training and experience provide a chary tolerance of biases of thought and indeed of easy acceptance of measures and methods of how aid works even in light of current-day knowledge of the forces of economic and social change in poor lands. Riddell is hard both on critics whose arguments condemn foreign aid as irrelevant—even harmful—for achievement of its goals and on those who readily extol foreign aid as fundamental to those ends. He is receptive to humanitarian considerations of a moral obligation on the part of rich nations to provide aid because the poor need help and even because of previous injustices. He opts for the middle ground. There have been positive achievements on both goals; developing lands are better off with foreign aid than without it. As readers might anticipate, the book ends with advocacy of higher levels of foreign aid, with relative increases for low-income groups of people and countries. He also argues for improved effectiveness: greater gains per unit of aid. In this, Riddell places heavy emphasis on the importance in recipient lands of a political and institutional environment with socioeconomic policies that foster modern development. This theme also emerges increasingly in current development literature. It recognizes the human quality input requirements of productive resource use. Meeting these involves political, social, cultural, and moral adjustments in the recipient nation's society. These can be made only through committed national leaders and over very long time periods.

The world's developed industrial nations initiated the modern epoch of growth some centuries back with the Industrial Revolution. In 1800, perhaps 20 percent of the world's population lived in these countries; they had initiated their social revolutions over earlier centuries. Only Japan has yet joined this group of rich nations. Its Meiji Revolution in the 1860s marked the emergence from feudalism to modern industrialism in a process of dramatic social and economic change spanning nearly a century. Government's role was primary and the process benefited from Japan's widespread education, common language, and Confucian faith.

There is little evidence of comparable beginnings toward higher productivity in today's low-income lands. Taken together, these countries, with an expanding bulk of the world's people, are still generating per capita incomes that are low and declining relative to those in the developed nations taken together. Riddell's study provides a valuable and comprehensive interpretation of the foreign aid record. His concluding observations on the necessity for improving effectiveness serve to underscore the formidable and relatively untouched tasks ahead.

WILFRED MALENBAUM
University of Pennsylvania
Philadelphia

OTHER BOOKS

ABEL, ELIE. *Leaking: Who Does It? Who Benefits? At What Cost?* Pp. vii, 75. New York: Priority Press, 1987. $18.95. Paperbound, $7.95.

ABRAHAM, KENNETH S. *Distributing Risk: Insurance, Legal Theory, and Public Policy.* Pp. 254. New Haven, CT: Yale University Press, 1986. $25.00.

AFANASYEV, V. G. *Historical Materialism.* Pp. vii, 180. New York: International Publishers, 1987. Paperbound, $4.95.

AFANASYEV, VICTOR G. *Dialectical Materialism.* Pp. vii, 152. New York: International Publishers, 1987. Paperbound, $4.25.

AHRARI, MOHAMMED E. *Ethnic Groups and U.S. Foreign Policy.* Pp. xxi, 178. Westport, CT: Greenwood Press, 1987. $35.00.

ARENDELL, TERRY. *Mothers and Divorce: Legal, Economic, and Social Dilemmas.* Pp. xiv, 221. Berkeley: University of California Press, 1986. No price.

ARROW, KENNETH J. and HERVE RAYNAUD. *Social Choice and Multi-Criterion Decision-Making.* Pp. vii, 127. Cambridge: MIT Press, 1986. $15.00.

BAHBAH, BISHARA. *Israel and Latin America: The Military Connection.* Pp. xvi, 210. New York: St. Martin's Press, in association with Institute for Palestine Studies, Washington, DC, 1986. No price.

BALOYRA, ENRIQUE A., ed. *Comparing New Democracies: Transition and Consolidation in Mediterranean Europe and the Southern Cone.* Pp. xviii, 318. Boulder, CO: Westview Press, 1987. $37.50.

BANERJI, SANJUKTA. *Deferred Hopes: Blacks in Contemporary America.* Pp. ix, 399. New York: Advent Books, 1987. $40.00.

BANKS, ARTHUR S., ed. *Political Handbook of the World, 1987.* Pp. x, 850. Binghamton, NY: Research Foundation of the State University of New York, 1987. $84.95.

BARTHE, PETER S. *The Tragedy of Black Lung: Federal Compensation for Occupational Disease.* Pp. xi, 292. Kalamazoo, MI: W. E. Upjohn Institute for Employment Research, 1987. Paperbound, no price.

BATINSKI, MICHAEL C. *The New Jersey Assembly, 1738-1775: The Making of a Legislative Community.* Pp. xx, 310. Lanham, MD: University Press of America, 1987. No price.

BAUM, ANN TODD. *Komsomol Participation in the Soviet First Five-Year Plan.* Pp. 62. New York: St. Martin's Press, 1987. $25.00.

BEATTY, BESS. *A Revolution Gone Backward: The Black Response to National Politics, 1876-1896.* Pp. xii, 235. Westport, CT: Greenwood Press, 1987. $35.00.

BERGSTRAND, SIMON and RIGAS DOGANIS. *The Impact of Soviet Shipping.* Pp. 184. Winchester, MA: Allen & Unwin, 1987. $40.00.

BERRIDGE, G. R. *The Politics of the South African Run: European Shipping and Pretoria.* Pp. xiii, 254. New York: Oxford University Press, 1987. $54.00.

BIJKER, WIEBE E., THOMAS P. HUGHES, and TREVOR PINCH, eds. *The Social Construction of Technological Systems: New Directions in the Sociology and History of Technology.* Pp. xi, 405. Cambridge: MIT Press, 1987. $35.00.

BINNENDIJK, HANS, ed. *Authoritarian Regimes in Transition.* Pp. xxvi, 336. Washington, DC: Department of State, 1987. Paperbound, no price.

BLECHMAN, BARRY M. and EDWARD N. LUTTWAK, eds. *Global Security: A Review of Strategic and Economic Issues.* Pp. xiv, 258. Boulder, CO: Westview Press, 1987. Paperbound, $29.95.

BONANNO, ALESSANDRO. *Small Farms: Persistence with Legitimation.* Pp. xvi, 228. Boulder, CO: Westview Press, 1987. Paperbound, $27.50.

BOURDIEU, PETER. *Distinction: A Social Critique of the Judgement of Taste.* Translated by Richard Nice. Pp. xiv, 613. Cambridge, MA: Harvard University Press, 1984. $29.50. Paperbound, $12.95.

BRAMS, STEVEN J. *Superpower Games: Applying Game Theory to Superpower Conflict.* Pp. xvi, 176. New Haven, CT: Yale University Press, 1985. $22.50. Paperbound, $6.95.

BROH, C. ANTHONY. *A Horse of a Different Color: Television's Treatment of Jesse Jackson's 1984 Presidential Campaign.* Pp. xiv, 93. Washington, DC: Joint Center for Political Studies, 1987. Paperbound, $7.95.

BROWN, LESTER R. et al. *State of the World, 1987.* Pp. xvii, 268. New York: Norton, 1987. Paperbound, $9.95.

BRYANT, RALPH C. *International Financial Intermediation.* Pp. xi, 181. Washington, DC: Brookings Institution, 1987. $9.95. Paperbound, $7.95.

BUCHANAN, JAMES M. *Liberty, Market, and State: Political Economy in the 1980s.* Pp. ix, 278. New York: New York University Press, 1986. Distributed by Columbia University Press, New York. $45.00.

BURDICK, CHARLES B. *An American Island in Hitler's Reich: The Bad Nauheim Internment.* Pp. 120. Menlo Park, CA: Markgraf, 1987. $32.95. Paperbound, $18.95.

CELMER, MARC A. *Terrorism, U.S. Strategy, and Reagan Policies.* Pp. x, 132. Westport, CT: Greenwood Press, 1987. $29.95.

CHEE, CHAN HENG and OBAID UL HAQ, eds. *S. Rajaratnam: The Prophetic and the Political.* Pp. 540. New York: St. Martin's Press, 1987. $45.00.

CHI, WEN-SHUN. *Ideological Conflicts in Modern China: Democracy and Authoritarianism.* Pp. xiii, 372. New Brunswick, NJ: Transaction Books, 1986. $29.95.

CIOFFI-REVILLA, CLAUDIO et al., eds. *Communication and Interaction in Global Politics.* Pp. 268. Newbury Park, CA: Sage, 1987. $29.95.

CLARK, PAUL F., PETER GOTTLIEB, and DONALD KENNEDY, eds. *Forging a Union of Steel: Philip Murray, SWOC, and United Steelworkers.* Pp. vii, 153. Ithaca, NY: ILR Press, 1987. $22.50. Paperbound, $8.95.

COKER, CHRISTOPHER. *NATO, the Warsaw Pact and Africa.* Pp. 302. New York: St. Martin's Press, 1985. $32.50.

COLE, GEORGE F., STANISLAW J. FRANKOWSKI, and MARC G. GERTZ, eds. *Major Criminal Justice Systems: A Comparative Survey.* 2nd ed. Newbury Park, CA: Sage, 1987. $22.95. Paperbound, $14.95.

CONLIN, JOSEPH R. *Bacon, Beans, and Galantines: Food and Foodways on the Western Frontier.* Pp. xii, 246. Reno: University of Nevada Press, 1986. No price.

CONRAD, JOHN P. *The Dangerous and the Endangered: The Dangerous Offender Project.* Pp. 153. Lexington, MA: D.C. Heath, Lexington Books, 1985. $19.00.

DAILEY, LOUIS E., Jr. *Justice Off-Balance: Did Injustice Flourish When "Equal Justice under Law" Collapsed?* Pp. 331. New York: Vantage Press, 1987. $16.95.

DANIELIAN, RONALD L. and STEPHEN E. THOMSEN. *The Forgotten Deficit: America's Addiction to Foreign Capital.* Pp. xii, 84. Boulder, CO: Westview Press, 1987. Paperbound, $11.85.

DANZIGER, JAMES N. and KENNETH L. KRAEMER. *People and Computers: The Impacts of Computing on End Users in Organizations.* Pp. xv, 248. New York: Columbia University Press, 1986. $32.50.

DEMOS, JOHN. *Past, Present, and Personal: The Family and the Life Course in American History.* Pp. xiii, 215. New York: Oxford University Press, 1986. $17.95.

DENG, FRANCIS MADING. *Tradition and Modernization: A Challenge for Law among the Dinka of the Sudan.* Pp. xlvii, 430. New Haven, CT: Yale University Press, 1987. Paperbound, $13.95.

DODER, DUSKO. *Shadows and Whispers: Power Politics Inside the Kremlin from Brezhnev to Gorbachev.* Pp. xi, 346. New York: Penguin Books, 1988. Paperbound, $7.95.

DOIG, JAMESON W. and ERWIN C. HARGROVE, eds. *Leadership and Innovation: A Biographical Perspective on Entrepreneurs in Government.* Pp. xii, 459. Baltimore, MD: Johns Hopkins University Press, 1987. $39.50.

DRENNAN, MATTHEW P. *Modeling Metropolitan Economies for Forecasting and Policy Analysis.* Pp. 242. New York: Columbia University Press, 1985. $40.00.

DUFFY, BERNARD K. and HALFORD R. RYAN, eds. *American Orators before 1900: Critical Studies and Sources.* Pp. xix, 481. Westport, CT: Greenwood Press, 1987. $75.00.

DUPRÉ, JOHN, ed. *The Latest and the Best: Essays on Evolution and Optimality.* Pp. xii, 359. Cambridge: MIT Press, 1987. $27.50.

DUTTON, WILLIAM H. et al., eds. *Wired Cities: Shaping the Future of Communications.* Pp. 492. Boston: G. K. Hall, 1987. Paperbound, $29.95.

EDELMAN, MARIAN WRIGHT. *Families in Peril: An Agenda for Social Change.* Pp. 127. Cambridge, MA: Harvard University Press, 1987. $15.00.

EHRLICH, ROBERT, ed. *Perspectives on Nuclear War and Peace Education.* Pp. vii, 242. Westport, CT: Greenwood Press, 1987. $37.95.

EISENHOWER, DAVID and JOHN MURRAY. *Warwords: U.S. Militarism, the Catholic Right and the "Bulgarian Connection."* Pp. 137. New York: International Publishers, 1986. Paperbound, no price.

EISENSTADT, S. N. *European Civilization in a Comparative Perspective.* Pp. 162. Oslo: Norwegian University Press, 1987. Distributed by Oxford University Press, New York. $45.00.

ELKIN, STEPHEN L. *City and Regime in the American Republic.* Pp. xi, 220. Chicago: University of Chicago Press, 1987. $35.00. Paperbound, $11.95.

EL-NAGGAR, SAID, ed. *Adjustment Policies and Development Strategies in the Arab World.* Pp. ix, 200. Washington, DC: International Monetary Fund, 1987. Paperbound, $12.00.

ENGLAND, PAULA and GEORGE FARKAS. *Households, Employment, and Gender: A Social, Economic and Demographic View.* Pp. 237. Hawthorne, NY: Aldine, 1986. Paperbound, no price.

ERIKSON, ROBERT and RUNE ABERG, eds. *Welfare in Transition: A Survey of Living Conditions in Sweden, 1968-1981.* Pp. xvii, 297. New York: Oxford University Press, 1987. $55.00.

Europe: Dream—Adventure—Reality. Pp. 262. Westport, CT: Greenwood Press, 1987. $65.00.

FERRAROTTI, FRANCO. *Five Scenarios for the Year 2000.* Pp. x, 135. Westport, CT: Greenwood Press, 1986. $29.95.

FEUER, LEWIS S. *Spinoza and the Rise of Liberalism.* Pp. xxx, 323. New Brunswick, NJ: Transaction Books, 1987. Paperbound, $19.95.

FILLER, LOUIS. *Crusade against Slavery: Friends, Foes and Reforms, 1820-1860.* Pp. 389. Algonac, MI: Reference Publications, 1986. $24.95. Paperbound, $12.95.

FISHER, JOSEPH and RICHARD T. MAYER, eds. *Virginia Alternatives for the 1990s: Selected Issues in Public Policy.* Pp. iv, 183. Fairfax, VA: George Mason University Press, 1987. Distributed by University Publishing Associates, Lanham, MD. $24.50. Paperbound, $10.95.

FLATHMAN, RICHARD E. *The Philosophy and Politics of Freedom.* Pp. 360. Chicago: University of Chicago Press, 1987. $42.50. Paperbound, $16.95.

FONER, NANCY, ed. *New Immigrants in New York.* Pp. ix, 318. New York: Columbia University Press, 1987. $27.50.

FOWLER, MARIAN. *Below the Peacock Fan: First Ladies of the Raj.* Pp. 337. New York: Viking, 1987. $19.95.

FREEDMAN, BEN. *To Be or Not to Be Human: The Traits of Human Nature.* Pp. ix, 525. New York: Vantage Press, 1987. $20.00.

FROST, ELLEN L. *For Richer, for Poorer: The New U.S.-Japan Relationship.* Pp. xiii, 199. Washington, DC: Council on Foreign Relations, 1987. Paperbound, no price.

FRUNDT, HENRY J. *Refreshing Pauses: Coca-Cola and Human Rights in Guatemala.* Pp. xvi, 269. New York: Praeger, 1987. $34.95.

GARNETT, JOHN C. *Commonsense and the Theory of International Politics.* Pp.

153. Albany: State University of New York Press, 1984. $29.50. Paperbound, $9.95.

GARTHOFF, RAYMOND L. *Policy versus the Law: The Reinterpretation of the ABM Treaty.* Pp. x, 117. Washington, DC: Brookings Institution, 1987. Paperbound, $8.95.

GASTIL, RAYMOND D. *Freedom in the World: Political Rights and Civil Liberties, 1986-1987.* Pp. 411. Westport, CT: Greenwood Press, 1987. $35.00.

GIRARD, RENÉ. *Things Hidden since the Foundation of the World.* Translated by Stephen Bann and Michael Metteer. Pp. 469. Stanford, CA: Stanford University Press, 1987. $30.00.

GLASSMAN, RONALD M., WILLIAM H. SWATOS, Jr., and PAUL L. ROSEN, eds. *Bureaucracy against Democracy and Socialism.* Pp. xi, 231. Westport, CT: Greenwood Press, 1987. $39.95.

GOLDBERG, ELLEN S. and DAN HAENDEL. *On Edge: International Banking and Country Risk.* Pp. 116. New York: Praeger, 1987. $32.95.

GOLDWIN, ROBERT A. and ART KAUFMAN, eds. *How Does the Constitution Protect Religious Freedom?* Pp. xv, 175. Washington, DC: American Enterprise Institute for Public Policy Research, 1987. $24.75. Paperbound, $12.50.

GOMES, GUSTAVO MAIA. *The Roots of State Intervention in the Brazilian Economy.* Pp. xvii, 376. New York: Praeger, 1986. $42.95.

GOTTFREDSON, STEPHEN D. and SEAN McCONVILLE, eds. *America's Correctional Crisis: Prison Populations and Public Policy.* Pp. vi, 260. Westport, CT: Greenwood Press, 1987. $37.95.

GRAHAM, LAWRENCE S. and MARIA K. CIECHOCIŃSKA, eds. *The Polish Dilemma: Views from Within.* Pp. xi, 258. Boulder, CO: Westview Press, 1987. $34.50.

GRINDLE, MERILEE S. *State and Countryside: Development Policy and Agrarian Politics in Latin America.* Pp. xiii, 255. Baltimore, MD: Johns Hopkins University Press, 1986. $25.00. Paperbound, $11.95.

GRUNFELD, A. TOM. *The Making of Modern Tibet.* Pp. 276. Armonk, NY: M. E. Sharpe, 1987. $27.50.

GUESS, GEORGE M. *The Politics of United States Foreign Aid.* Pp. 297. New York: St. Martin's Press, 1987. $32.50.

GUMPERT, GARY. *Talking Tombstones & Other Tales of the Media Age.* Pp. 205. New York: Oxford University Press, 1987. $17.95.

GURR, TED ROBERT and DESMOND S. KING. *The State and the City.* Pp. ix, 242. Chicago: University of Chicago Press, 1987. $35.00. Paperbound, $14.95.

GUYER, JANE L. *Feeding African Cities: Studies in Regional Social History.* Pp. x, 249. Bloomington: Indiana University Press, 1987. $29.95.

HADDAD, YVONNE YAZBECK and ADAIR T. LUMMIS. *Islamic Values in the United States.* Pp. viii, 196. New York: Oxford University Press, 1987. Paperbound, $24.95.

HAHNER, JUNE E. *Poverty and Politics: The Urban Poor in Brazil, 1870-1920.* Pp. xvi, 415. Albuquerque: University of New Mexico Press, 1986. No price.

HAMM, MICHAEL F. *The City in Late Imperial Russia.* Pp. 372. Bloomington: Indiana University Press, 1986. $27.50.

HANRIEDER, WOLFRAM F., ed. *Arms Control, the FRG, and the Future of East-West Relations.* Pp. xi, 138. Boulder, CO: Westview Press, 1987. $28.50.

HANRIEDER, WOLFRAM F., ed. *Global Peace and Security: Trends and Challenges.* Pp. x, 223. Boulder, CO: Westview Press, 1987. $30.00.

HARDING, SANDRA, ed. *Feminism and Methodology.* Pp. ix, 193. Bloomington: Indiana University Press, 1987. $29.95. Paperbound, $10.95.

HAZAN, BARUCH. *From Brezhnev to Gorbachev: Infighting in the Kremlin.* Pp. xii, 260. Boulder, CO: Westview Press, 1987. $34.95.

HELLER, AGNES. *Beyond Justice.* Pp. vi, 346. New York: Basil Blackwell, 1987. $39.95.

HENDRICK, CLYDE, ed. *Group Processes.* Pp. 294. Newbury Park, CA: Sage, 1987. Paperbound, no price.

HENDRICK, CLYDE, ed. *Group Processes and Intergroup Relations.* Pp. 256. Newbury Park, CA: Sage, 1987. Paperbound, no price.

HOFFMAN, JOHN. *The Gramscian Challenge: Coercion and Consent in Marxist Political Theory.* Pp. 230. New York: Basil Blackwell, 1984. Paperbound, $12.95.

HOVANNISIAN, RICHARD G. *The Armenian Genocide in Perspective.* Pp. 215. New Brunswick, NJ: Transaction Books, 1986. $29.95. Paperbound, $14.95.

HUSTON, JAMES A. *One for All: NATO Strategy and Logistics through the Formative Period, 1949-69.* Pp. 332. Newark: University of Delaware Press, 1984. $38.50.

JACKSON, DOUGLAS N. and J. PHILLIPPE RUSHTON, eds. *Scientific Excellence: Origins and Assessment.* Pp. 381. Newbury Park, CA: Sage, 1987. $28.00.

JACOBSEN, CARL G., ed. *The Uncertain Course: New Weapons Strategies and Mind Sets.* Pp. xxiii, 349. New York: Oxford University Press, 1987. $64.00.

JENISTA, FRANK LAWRENCE. *The White Apos: American Governors on the Cordillera Central.* Pp. xii, 321. Quezon City, Philippines: New Day, 1987. Distributed by Cellar Book Shop, Detroit, MI. Paperbound, $14.00.

KAUFMANN, WILLIAM W. *A Thoroughly Efficient Navy.* Pp. xii, 131. Washington, DC: Brookings Institution, 1987. Paperbound, $8.95.

KELLY, ALFRED, ed. *The German Worker: Working Class Autobiographies from the Age of Industrialization.* Translated by Alfred Kelly. Pp. xiii, 438. Berkeley: University of California Press, 1987. $45.00. Paperbound, $12.95.

KOZAR, PAUL MICHAEL. *The Politics of Deterrence: American and Soviet Defense Policies Compared, 1960-1964.* Pp. vi, 169. Jefferson, NC: McFarland, 1987. $24.95.

KURLAND, PHILIP B. and RALPH LERNER, eds. *The Founders' Constitution: Major Themes.* Pp. xiii, 713. Chicago: University of Chicago Press, 1987. $49.95. Paperbound, $24.50.

LAMPTON, DAVID M., ed. *Policy Implementation in Post-Mao China.* Pp. xii, 439. Berkeley: University of California Press, 1987. $48.00.

LANE, JAN-ERIK. *State and Market: The Politics of the Public and the Private.* Pp. 302. Newbury Park, CA: Sage, 1985. $40.00. Paperbound, $16.00.

LAQUEUR, WALTER and BRAD ROBERTS, eds. *America in the World, 1962-1987: A Strategic and Political Reader.* Pp. xii, 468. New York: St. Martin's Press, 1987. No price.

LAZERE, DONALD. *American Media and Mass Culture: Left Perspectives.* Pp. xii, 618. Berkeley: University of California Press, 1987. $48.00. Paperbound, $15.95.

LEWIS, GORDON K. *Grenada: The Jewel Despoiled.* Pp. x, 239. Baltimore, MD: Johns Hopkins University Press, 1987. Paperbound, $25.00.

LIPSON, LEON and STANTON WHEELER, eds. *Law and the Social Sciences.* Pp. viii, 740. New York: Russell Sage Foundation, 1987. Distributed by Basic Books, New York. $65.00.

LITTLE, PETER D. et al., eds. *Lands at Risk in the Third World: Local-Level Perspectives.* Boulder, CO: Westview Press, 1987. Paperbound, $24.95.

LODGARD, SVERRE and KARL BIRNBAUM, eds. *Overcoming Threats to Europe: A New Deal for Confidence and Security.* Pp. ix, 235. New York: Oxford University Press, 1987. No price.

LOVENDUSKI, JONI and JEAN WOODALL. *Politics and Society in Eastern Europe.* Pp. xiii, 474. Bloomington: Indiana University Press, 1987. $45.00. Paperbound, $14.50.

MACLEAN, HELENE. *Caring for Your Parents: A Sourcebook of Options and Solutions for Both Generations.* Pp. 370. New York: Doubleday, 1987. Paperbound, $12.95.

MACPHERSON, STEWART and JAMES MIDGLY. *Comparative Social Policy and the Third World: Studies in International Social Policy and Welfare.* Pp. 228. New York: St. Martin's Press, 1987. $35.00.

MADDEN, FREDERICK, ed., with David Fieldhouse. *Imperial Reconstruction, 1763-1840: The Evolution of Alternative Systems of Colonial Government.* Vol. 3, *Select Documents on the Constitutional History of the British Empire and Commonwealth.* Pp. xl, 888. Westport, CT: Greenwood Press, 1987. $95.00.

MAJOR, JOHN S. and ANTHONY J. KANE, eds. *China Briefing, 1987.* Pp. ix, 197. Boulder, CO: Westview Press, 1987. $28.85. Paperbound, $13.85.

MARCUS, ALFRED A., ALLEN M. KAUFMAN, and DAVID R. BEAM, eds. *Business Strategy and Public Policy: Perspectives from Industry and Academia.* Pp. xv, 323. Westport, CT: Greenwood Press, Quorum Books, 1987. $49.95.

MARER, PAUL and WŁODZIMIERZ SIWIŃSKI, eds. *Creditworthiness and Reform in Poland: Western and Polish Perspectives.* Pp. xxiii, 348. Bloomington: Indiana University Press, 1987. $37.50. Paperbound, $25.00.

MARSHALL, SHERRIN. *The Dutch Gentry, 1500-1650: Family, Faith, and Fortune.* Pp. 225. Westport, CT: Greenwood Press, 1987. $35.00.

MARTIRENA-MANTEL, ANA MARIA, ed. *Eternal Debt, Savings, and Growth in Latin America.* Pp. xv, 207. Washington, DC: International Monetary Fund, 1987. $12.00.

McCARTHY, PATRICK, ed. *The French Socialists in Power, 1981-1986.* Pp. xii, 212. Westport, CT: Greenwood Press, 1987. $39.95.

McDONALD, JOHN W., Jr. and DIANE B. BENDAHMANE, eds. *Conflict Resolution: Track Two Diplomacy.* Pp. 87. Washington, DC: Department of State, 1987. Paperbound, no price.

MEYER, JACK ALLEN, ed. *An Annotated Bibliography of the Napoleonic Era: Recent Publications, 1945-1985.* Pp. xvii, 288. Westport, CT: Greenwood Press, 1987. $39.95.

MILES, WILLIAM. *Elections in Nigeria: A Grassroots Perspective.* Pp. 168. Boulder, CO: Lynne Rienner, 1987. No price.

MILLER, DAVID et al., eds. *The Blackwell Encyclopedia of Political Thought.* Pp. xiii, 570. New York: Basil Blackwell, 1987. $60.00.

MOMEN, MOOJAN. *An Introduction to Shi'i Islam.* Pp. 397. New Haven, CT: Yale University Press, 1987. $15.95.

MORELAND, LAURENCE, ROBERT P. STEED, and TOD A. BAKER, eds. *Blacks in Southern Politics.* Pp. ix, 305. New York: Praeger, 1987. $42.95.

MOSQUEDA, LAWRENCE J. *Chicanos, Catholicism and Political Ideology.* Pp. 219. Lanham, MD: University Press of America, 1986. $24.50. Paperbound, $12.75.

MOSS, ARMAND. *Disinformation, Misinformation, and the "Conspiracy" to Kill JFK Exposed.* Pp. viii, 219. Hamden, CT: Archon Books, 1987. $22.50.

MULLER, KLAUS-JURGEN. *The Army, Politics, and Society in Germany, 1933-1945.* Pp. ix, 122. New York: St. Martin's Press, 1987. $25.00.

NAGEL, JACK H. *Participation.* Pp. 195. Englewood Cliffs, NJ: Prentice-Hall, 1987. Paperbound, no price.

NEVINS, JANE. *Turning 200: A Bicentennial History of the Rise of the American Republic.* Pp. xii, 395. New York: Richardson & Steirman, 1987. $21.95.

NEWMAN, EDGAR LEON. *Historical Dictionary of France from the 1815 Restoration to the Second Empire.* 2 vols. Pp. xvii, 1215. Westport, CT: Greenwood Press, 1987. $135.00.

NEWSOM, DAVID D. *The Soviet Brigade in Cuba: A Study in Political Diplomacy.* Pp. xvi, 122. Bloomington: Indiana University Press, 1987. $25.00. Paperbound, $7.95.

NICHOLS, W. GARY and MILTON L. BOYKIN, eds. *Arms Control and Nuclear Weapons: U.S. Policies and the National Interest.* Pp. xii, 135. Westport, CT: Greenwood Press, 1987. $29.95.

NIISEKI, KINYA, ed. *The Soviet Union in*

Transition. Pp. ix, 243. Boulder, CO: Westview Press, 1987. $34.50.

NISBET, ROBERT. *The Making of Modern Society.* Pp. v, 215. New York: New York University Press, 1987. Distributed by Columbia University Press, New York. $35.00.

NORMAN, RICHARD. *Free and Equal: A Philosophical Examination of Political Values.* Pp. 178. New York: Oxford University Press, 1987. $37.00. Paperbound, $12.95.

O'BRIEN, WILLIAM V. and JOHN LANGAN. *The Nuclear Dilemma & the Just War Tradition.* Pp. 260. Lexington, MA: D. C. Heath, 1986. $25.00.

ORREN, GARY R. and NELSON W. POLSBY, eds. *Media and Momentum: The New Hampshire Primary and Nomination Politics.* Pp. iv, 200. Chatham, NJ: Chatham House, 1987. $20.00. Paperbound, $11.95.

OSTROM, VINCENT. *The Political Theory of a Compound Republic: Designing the American Experiment.* Pp. 240. Lincoln: University of Nebraska Press, 1987. $22.50. Paperbound, $8.95.

PERKINS, DWIGHT H. *China: Asia's Next Economic Giant.* Pp. x, 98. Seattle: University of Washington Press, 1986. $12.95.

PERSELL, CAROLINE HODGES. *Understanding Society: An Introduction to Sociology.* Pp. xxv, 652. New York: Harper & Row, 1987. No price.

PETERS, A. R. *Anthony Eden at the Foreign Office 1931-1938.* Pp. 402. New York: St. Martin's Press, 1987. $39.95.

PIERRE, ANDREW J., ed. *A High Technology Gap: Europe, America, and Japan.* Pp. xii, 114. New York: Council on Foreign Relations, 1987. Paperbound, $6.95.

PITT, DAVID and THOMAS G. WEISS. *The Nature of United Nations Bureaucracies.* Pp. 199. Boulder, CO: Westview Press, 1986. $28.50.

PORTER, A. N. and STOCKWELL, A. J. *British Imperial Policy and Decolonization, 1938-1964.* Vol. 1, *1938-1951.* Pp. xvii, 403. New York: St. Martin's Press, 1987. $39.95.

RAINO, KULLERVO. *Stochastic Field Theory of Behavior.* Pp. 250. Helsinki: Finnish Society of Sciences and Letters, 1986. Paperbound, no price.

RANGEL, CARLOS. *The Latin Americans: Their Love-Hate Relationship with the United States.* Pp. xvii, 312. New Brunswick, NJ: Transaction Books, 1987. Paperbound, $12.95.

RANKIN, MARY BACKUS. *Elite Activism and Political Transformation in China: Zhejiang Province, 1865-1911.* Pp. 427. Stanford, CA: Stanford University Press, 1986. $39.50.

RECTOR, ROBERT and MICHAEL SANERA, eds. *Steering the Elephant: How Washington Works.* Pp. 356. New York: Universe, 1987. $24.95.

REEVE, ANDREW, ed. *Modern Theories of Exploitation.* Pp. 209. Newbury Park, CA: Sage, 1987. $39.95. Paperbound, $17.95.

ROBINSON, DEREK. *Monetarism and the Labour Market.* Pp. xvi, 499. New York: Oxford University Press, 1986. $49.95.

RONNING, C. NEALE and ALBERT P. VANNUCCI, eds. *Ambassadors in Foreign Policy: The Influence on U.S.-Latin American Policy.* Pp. xii, 154. New York: Praeger, 1987. $36.95.

SAMUELS, WARREN J. and ARTHUR S. MILLER, eds. *Corporations and Society: Power and Responsibility.* Pp. xv, 328. Westport, CT: Greenwood Press, 1987. $45.00.

SANG-WOO, RHEE, ed. *Korean Unification: Source Materials with an Introduction.* Vol. 3. Pp. xxxv, 582. Seoul: Research Center for Peace and Unification of Korea, 1986. No price.

SAVAS, E. S. *Privatization: The Key to Better Government.* Pp. xi, 308. Chatham, NJ: Chatham House, 1987. $25.00. Paperbound, $14.95.

SCALAPINO, ROBERT A. *Major Power Relations in Northeast Asia.* Pp. 70. Lanham, MD: University Press of America, 1987. $15.50. Paperbound, $4.75.

SCHWARZMANTEL, JOHN. *Structures of Power: An Introduction to Politics.*

Pp. ix, 270. New York: St. Martin's Press, 1987. $35.00.

SCHWOK, RENE. *Interprétations de la politique étrangère de Hitler: Une analyse de l'historiographie.* Pp. 217. Paris: Presses universitaires de France, 1987. Paperbound, no price.

SEYMOUR, JAMES D. *China's Satellite Parties.* Pp. 149. Armonk, NY: M. E. Sharpe, 1987. $25.00.

SHEPHERD, WILLIAM G. *The Ultimate Deterrent: Foundations of US-USSR Security under Stable Competition.* Pp. x, 137. New York: Praeger, 1986. $29.95.

SHIELDS, ART. *On the Battle Lines, 1919-1939.* Pp. 278. New York: International Publishers, 1987. $14.00. Paperbound, $6.95.

SLAVIN, SARAH, ed. *The Politics of Professionalism, Opportunity, Employment, and Gender.* Pp. 110. New York: Haworth Press, 1987. $19.95.

SPITZER, ROBERT J. *The Right to Life and Third Party Politics.* Pp. xii, 154. Westport, CT: Greenwood Press, 1987. $29.95.

STALSON, HELENA. *Intellectual Property Rights and U.S. Competitiveness in Trade.* Pp. x, 106. Washington, DC: National Planning Association, 1987. Paperbound, $15.00.

STANLEY, HAROLD W. *Voter Mobilization and the Politics of Race: The South and Universal Suffrage, 1952-1984.* Pp. xv, 185. New York: Praeger, 1987. $35.00.

STERN, ROBERT M. *U.S. Trade Policies in a Changing World Economy.* Pp. 437. Cambridge: MIT Press, 1987. $25.00.

STOVER, ERIC. *The Open Secret: Torture and the Medical Profession in Chile.* Pp. vi, 80. Washington, DC: American Association for the Advancement of Science, 1987. Paperbound, no price.

STRAUSS, LEO and JOSEPH CROPSEY, eds. *History of Political Philosophy.* 3rd ed. Pp. xiv, 966. Chicago: University of Chicago Press, 1987. $75.00. Paperbound, $19.95.

THAKUR, RAMESH and CARLYLE A. THAYER, eds. *The Soviet Union as an Asian Pacific Power: Implications of Gorbachev's 1986 Vladivostok Initiative.* Pp. vi, 236. Boulder, CO: Westview Press, 1987. Paperbound, $31.30.

THURSTON, ROBERT W. *Liberal City, Conservative State: Moscow and Russia's Urban Crisis, 1906-1914.* Pp. xi, 266. New York: Oxford University Press, 1987. $32.00.

TOBIN, MAURICE. *Hidden Power: The Seniority System and Other Customs of Congress.* Pp. 134. Westport, CT: Greenwood Press, 1986. $27.95.

VAKSMAN, FABIAN. *Ideological Struggle: A Study in the Principles of Operation of the Soviet Political Mechanism.* Pp. ix, 215. Lanham, MD: University Press of America, 1987. No price.

VALENTA, JIRI and ESPERANZA DURAN. *Conflict in Nicaragua: A Multidimensional Perspective.* Pp. 440. Winchester, MA: Allen & Unwin, 1987. $45.00. Paperbound, $19.95.

VAN NIEKERK, BAREND. *The Cloistered Virtue: Freedom of Speech and the Administration of Justice in the Western World.* Pp. 399. Westport, CT: Praeger, 1987. $49.95.

VAN ZYL SLABBERT, FREDERIK. *The Last White Parliament: The Struggle for South Africa by the Leader of the White Opposition.* Pp. 203. New York: St. Martin's Press, 1987. $24.95.

VICKERS, GEOFFREY. *Policymaking, Communication, and Social Learning.* Edited by Guy Adams, John Forester, and Bayard L. Catron. Pp. xiv, 202. New Brunswick, NJ: Transaction Books, 1987. $34.95.

VISCUSI, W. KIP and WESLEY A. MAGAT. *Learning about Risk: Consumer and Worker Responses to Hazard Information.* Pp. xi, 197. Cambridge, MA: Harvard University Press, 1987. $27.50.

WEINBERG, ALVIN M. and JACK N. BARKENBUS, eds. *Strategic Defenses and Arms Control.* Pp. viii, 263. New York: Paragon House, 1987. $24.95. Paperbound, $12.95.

WHEATON, ELIZABETH. *Codename Green-*

kil: *The 1979 Greensboro Killings.* Pp. x, 328. Athens: University of Georgia Press, 1987. $24.95.

WHICKER, MARCIA LYNN, RUTH ANN STRICKLAND, and RAYMOND A. MOORE. *The Constitution under Pressure: A Time for Change.* Pp. xii, 223. New York: Praeger, 1987. $38.95. Paperbound, $14.95.

WHITE, LANDEG. *Magomero: Portrait of an African Village.* Pp. xii, 271. New York: Cambridge University Press, 1987. No price.

WILKS, STEPHEN and MAURICE WRIGHT, eds. *Comparative Government-Industry Relations: Western Europe, the United States, and Japan.* Pp. xiii, 321. New York: Oxford University Press, 1987. $55.00.

WILLEMS, EMILIO. *A Way of Life and Death: Three Centuries of Prussian-German Militarism, an Anthropological Approach.* Pp. 219. Champaign: University of Illinois Press, 1986. $12.95.

WILLETT, T. C. *A Heritage at Risk: The Canadian Militia as a Social Institution.* Pp. xviii, 269. Boulder, CO: Westview Press, 1987. Paperbound, $27.50.

WINHAM, GILBERT R. *International Trade and the Tokyo Round Negotiation.* Pp. xiv, 449. Princeton, NJ: Princeton University Press, 1987. $45.00. Paperbound, $13.50.

WISHMAN, SEYMOUR. *Anatomy of a Jury: The Inside Story of How 12 Ordinary People Decide the Fate of an Accused Murderer.* Pp. 313. New York: Penguin, 1986. Paperbound, $6.95.

WOLDRING, HENK E.S. *Karl Manheim: The Development of His Thought.* Pp. xi, 450. New York: St. Martin's Press, 1987. $37.50.

WORLD COMMISSION ON ENVIRONMENT AND DEVELOPMENT. *Our Common Future.* Pp. xv, 383. New York: Oxford University Press, 1987. Paperbound, $9.95.

WORTZEL, LARRY M. *Class in China: Stratification in a Classless Society.* Pp. xi, 171. Westport, CT: Greenwood Press, 1987. $32.95.

WRIGHTSMAN, LAWRENCE S., SAUL M. KASSIN, and CYNTHIA E. WILLIS, eds. *In the Jury Box: Controversies in the Courtroom.* Pp. 260. Newbury Park, CA: Sage, 1987. $35.00. Paperbound, $17.95.

WRIGHTSMAN, LAWRENCE, S., SAUL M. KASSIN, and CYNTHIA E. WILLIS, eds. *On the Witness Stand: Controversies in the Courtroom.* Pp. 312. Newbury Park, CA: Sage, 1987. $37.50. Paperbound, $19.95.

YAMAMURA, KOZO and YASUKICHI YASUBA, eds. *The Political Economy of Japan.* Vol. 1, *The Domestic Transformation.* Pp. xxvi, 666. Stanford, CA: Stanford University Press, 1987. $37.50. Paperbound, $12.95.

YARBROUGH, TINSLEY E. *A Passion for Justice: J. Waties Waring and Civil Rights.* Pp. xii, 282. New York: Oxford University Press, 1987. $32.50.

YUAN, GAO. *Born Red: A Chronicle of the Cultural Revolution.* Pp. 380. Stanford, CA: Stanford University Press, 1987. $39.50. Paperbound, $7.95.

INDEX

Adams Express, 16
Air Transport Association, 66
Alarms (security industry), 24-27
American District Telegraph Company, 26
American Express Company, 16
American Society for Industrial Security (ASIS), 103, 104-5, 106
Antitrust Division, Department of Justice, 26, 28
Armored car industry, 27-29
Assumption of risk, doctrine of, 95
Auditors, and detection of fraud, 78-80

Bahamas, police services, 116
Bensinger, Peter B., 10
BENSINGER, PETER B., Drug Testing in the Workplace, 43-50
Biological screening of employees, 41
Brink, Washington Perry, 27
Brink's, 27-28

CAN POLICE SERVICES BE PRIVATIZED? Philip E. Fixler, Jr. and Robert W. Poole, Jr., 108-18
Carnegie, Phipps Steel Company, 30
Certified Protection Professional program, 104-6
Civil aviation, security measures in, 60-69
CIVIL AVIATION: TARGET FOR TERRORISM, William A. Crenshaw, 60-69
Civil liberty, 80-81
Common law, 12
Commonwealth Edison, 50
Company loyalty, 56, 58
Computers
  fradulent use of, 53, 71
  hazards to, 71-72
  protection of, 72-78, 81-82
Contributory negligence, doctrine of, 94
Crenshaw, William A., 10
CRENSHAW, WILLIAM A., Civil Aviation: Target for Terrorism, 60-69
Criminal acts by third parties, 95-97
Criminal records, 37-38, 87-89
Criscuoli, Ernest J., Jr., 10
CRISCUOLI, ERNEST J., Jr., The Time Has Come to Acknowledge Security as a Profession, 98-107
Cybernex, 55

DEVELOPMENT OF THE U.S. SECURITY INDUSTRY, THE, Robert D. McCrie, 23-33

Drug testing, 39, 43-50, 86-87
  court decisions, 47-48
  legislation, 48
DRUG TESTING IN THE WORKPLACE, Peter B. Bensinger, 43-50

Employee theft, 30, 51-59
EMPLOYEE THEFT: A $40 BILLION INDUSTRY, Mark Lipman and W. R. McGraw, 51-59
Employment history, 38-39, 85
England, private security in, 13, 14
Ethics in business, 55-56
Explosives detection, 66-67
Express companies, 16, 21, 27-29

Fargo, Walter, 16
Federal Bureau of Investigation, 88
Fixler, Philip E., Jr., 10
FIXLER, PHILIP E., Jr. and ROBERT W. POOLE, Jr., Can Police Services Be Privatized? 108-18
Fleming, James Douglas, 26
Frick, Henry Clay, 19, 30

Gendarmerie, 13
Genetic screening of employees, 40-41
Grinell, 26

Harris, Gary K., 10
HARRIS, GARY K., see SCHILLER, JONATHAN D., coauthor
Holmes, Edwin, 24-25
Homestead Works, 19-20, 30

IBM, 55
Interstate Commerce Commission, 28
Interstate Identification Index, 88-89

Janissaries, 12

Kalamazoo, Michigan, police services, 113

Labor strikes, 19-20, 30
Law enforcement, 31, 103, 104
  agencies, 21
  see also Police services
Law Enforcement Assistance Administration, 31
Legal concerns with fraud, 80-81
LEGAL LIABILITY OF A PRIVATE SE-

INDEX

CURITY GUARD COMPANY FOR THE CRIMINAL ACTS OF THIRD PARTIES: AN OVERVIEW, THE, Jonathan D. Schiller and Gary K. Harris, 91-97
Lincoln, Abraham, 17-18
LIPMAN, IRA A., Personnel Selection in the Private Security Industry: More Than a Résumé, 83-90
LIPMAN, IRA A., Preface, 9-10
Lipman, Mark, 10
LIPMAN, MARK and W. R. McGRAW, Employee Theft: A $40 Billion Industry, 51-59
Lipson, Milton, 10
LIPSON, MILTON, Private Security: A Retrospective, 11-22

Mamlukes, 12
McCrie, Robert D., 10
McCRIE, ROBERT D., The Development of the U.S. Security Industry, 23-33
McGraw, W. R., 10
McGRAW, W. R., see LIPMAN, MARK, coauthor
Minnesota Multiphasic Personality Inventory, 39, 87
Molly Maguires, 19
Montclair, California, police services, 111

National Advisory Committee on Criminal Justice Standards and Goals, 31
Night watch, 14, 15
Nonmetallic weapons, 66

Oro Valley, Arizona, police services, 114
Overman, Robert W., 10
OVERMAN, ROBERT W., Personnel Selection in Private Industry: The Role of Security, 34, 42

Personnel selection, 34-42, 58, 68, 74, 83-90
PERSONNEL SELECTION IN PRIVATE INDUSTRY: THE ROLE OF SECURITY, Robert W. Overman, 34-42
PERSONNEL SELECTION IN THE PRIVATE SECURITY INDUSTRY: MORE THAN A RÉSUMÉ, Ira A. Lipman, 83-90
Pinkerton, Allan, 16-20, 21, 29-30
Police, public, 15
Police services, 108-18
Polygraph, 37-38, 39, 85-86, 90

Pomeranz, Felix, 10
POMERANZ, FELIX, Technological Security, 70-82
Poole, Robert W., Jr., 10
POOLE, ROBERT W., Jr., see FIXLER, PHILIP E., Jr., coauthor
Praetorian Guard, 12
Privacy, protection of, 35-36, 39-40, 41, 45, 47, 63, 80, 85, 90
Private goods, 109-10
Private justice system, 56-57
Private security industry, 11-22, 23-33, 39, 83-90
professionalism, 98-107
public policy, 32-33
PRIVATE SECURITY: A RETROSPECTIVE, Milton Lipson, 11-22
Privatization of police services, 110-18
Public goods, 109-10, 117

Railroads, security of, 16, 29, 116
Rand Report, The, 31
Reasonable care, standard of, 93-95, 97
Reminderville, Ohio, police services, 113-14
Risk management, 75-76
Rome, fall of, 12, 13

San Francisco, police services, 116-17
Savas, E. S., 109
Schiller, Jonathan D., 10
SCHILLER, JONATHAN D. and GARY K. HARRIS, The Legal Liability of a Private Security Guard Company for the Criminal Acts of Third Parties: An Overview, 91-97
Security guard industry, 29-32, 83-90

TECHNOLOGICAL SECURITY, Felix Pomeranz, 70-82
Terrorism, 60-69
TIME HAS COME TO ACKNOWLEDGE SECURITY AS A PROFESSION, THE, Ernest J. Criscuoli, Jr., 98-107

University of Southern California, campus police, 116

Wells Fargo and Company, 16
Wells, Henry, 16
West (U.S.), settlement of, 20-21
World War I, 21

NEW from Sage

IN THE JURY BOX
Controversies In The Courtroom
edited by LAWRENCE S. WRIGHTSMAN, *University of Kansas*
SAUL KASSIN, *Williams University*
& CYNTHIA WILLIS, *University of Kansas*

How well does the jury selection process "work"? Are jurors able to make unbiased judgments? Should complex cases be determined by an experienced judge rather than a set of novice jurors? Does it make any difference in verdicts if the size of the jury is reduced or the decision rule changed? Questions such as these have long fascinated and mystified the public, the news media, and legal scholars. **In the Jury Box** is a collection of articles from the accumulating professional literature that examines these and other challenging questions surrounding the jury process.

Designed for a broad interdisciplinary audience, **In the Jury Box** is appropriate for use as a text supplement in undergraduate psychology, law, and criminal justice courses.

CONTENTS: I. Jury Selection // II. The Biased Juror // III. Jury Competence // IV. Jury Size and Decision Rule

1987 (Autumn) / 256 pages (tent.) / $35.00 (c) / $19.95 (p)

ON THE WITNESS STAND
Controversies in the Courtroom
edited by LAWRENCE S. WRIGHTSMAN
& CYNTHIA E. WILLIS, *both at University of Kansas*
& SAUL KASSIN, *Williams College*

In a trial, whether it be criminal or civil, evidence can be introduced only through exhibits or the testimony of witnesses, and the witnesses are subject to the limitations and mistakes that are present in every human being. Recently, psychologists and other social scientists have begun to empirically evaluate questions such as: What is the impact of the testimony of eyewitnesses in court decisions? Should hypnosis be used with crime witnesses? What effect do expert witnesses have? Does the presence of television cameras affect the testimony of witnesses? **On the Witness Stand** is a carefully selected collection of articles reprinted from psychological journals relevant to the above questions. **On the Witness Stand** will interest not only students, teachers, and researchers in the social sciences, but also trial lawyers and judges as it examines, from a unique perspective, the role of witnesses in court.

CONTENTS: I. Lie Detection and Polygraph Testing // II. Refreshing Memory Through Hypnosis // III. Eyewitness Accuracy // IV. Expert Witness Testimony // V. Cameras in the Courtroom

1987 / 296 pages (tent.) / $37.50 (c) / $19.95 (p)

SAGE PUBLICATIONS, INC.
2111 West Hillcrest Drive,
Newbury Park, California 91320

SAGE PUBLICATIONS, INC.
275 South Beverly Drive,
Beverly Hills, California 90212

SAGE PUBLICATIONS LTD
28 Banner Street,
London EC1Y 8QE, England

SAGE PUBLICATIONS INDIA PVT LTD
M-32 Market, Greater Kailash I,
New Delhi 110 048 India

NEW from Sage

COMMUNITY CRIME PREVENTION
Does It Work?
edited by DENNIS P. ROSENBAUM, *Northwestern University*

Since the "war on crime" began in the mid-1960s many research efforts have been undertaken to probe the nature of this nation's serious crime problem and to propose ways of solving it. Criminologists, policymakers, and law enforcement officials have since recognized that community organizations can and must collectively participate with law enforcement officials in prevention activities. This volume represents one of the first major attempts to bring together nationally respected criminal justice scholars and their evaluations of crime prevention programs. These evaluations are presented in four groups: (a) citizens' initiatives to prevent residential crime; (b) police initiatives to prevent residential crime; (c) programs to prevent commercial crime; and (d) media strategies for influencing crime prevention attitudes and behavior. All address the fundamental question of whether citizen and police initiatives make any impact on crime. This volume is unique in that its contributors address a broad and varied audience, including practitioners, policymakers, and scholars interested in community crime prevention and evaluation research.

CONTENTS: Foreword // **I. The Evaluation Problem** // 1. The Problem of Crime Control D.P. ROSENBAUM / 2. Evaluation Research in Community Crime Prevention A.J. LURIGIO & D.P. ROSENBAUM // **II. Citizen Efforts to Prevent Residential Crime** // 3. Citywide Community Crime Prevention B. LINDSAY & D. McGILLIS / 4. Neighborhood-Based Anti-Burglary Strategies A.L. SCHNEIDER / 5. A Three-Pronged Effort to Reduce Crime and Fear of Crime F.J. FOWLER & T.W. MANGIONE / 6. Neighborhood-Based Crime Prevention D.P. ROSENBAUM, D.A. LEWIS & J.A. GRANT // **III. Innovations in Policing: A Return to the Neighborhood** // 7. Experimenting with Foot Patrol A.M. PATE / 8. Evaluating a Neighborhood Foot Patrol Program R.C. TRANJANOWICZ / 9. Storefront Police Offices M.A. WYCOFF & W.G. SKOGAN // **IV. Preventing Crime in and Around Commercial Establishments** // 10. Evaluating Crime Prevention Through Environmental Design P.J. LAVRAKAS & J.W. KUSHMUK / 11. The Commercial Security Field Test Program J.M. TIEN & M.F. CAHN // **V. The Media and Crime Prevention: Public Education and Persuasion** // 12. The "McGruff" National Media Campaign G.J. O'KEEFE / 13. Evaluating Police-Community Anti-Crime Newsletters P.L. LAVRAKAS // **VI. Summary and Critique** // 14. Community Crime Prevention R.K. YIN

Criminal Justice Systems Annual, Volume 22
1986 / 318 pages / $29.95 (c) / $14.95 (p)

SAGE PUBLICATIONS, INC.
2111 West Hillcrest Drive,
Newbury Park, California 91320

SAGE PUBLICATIONS, INC.
275 South Beverly Drive,
Beverly Hills, California 90212

SAGE PUBLICATIONS LTD
28 Banner Street,
London EC1Y 8QE, England

SAGE PUBLICATIONS INDIA PVT LTD
M-32 Market, Greater Kailash I,
New Delhi 110 048 India

NEW from Sage

HOUSE ARREST AND CORRECTIONAL POLICY
Doing Time at Home
by **RICHARD BALL**, *University of West Virginia,*
C. RONALD HUFF, *Ohio State University*
& **J. ROBERT LILLY**, *Northern Kentucky University*

"Too often technology-driven reforms offer us answers, but fail to tell us what the questions are. This useful little book helps us remember the questions! Amidst the commercial and moral entrepreneurs promising electronic salvation, it offers a welcome cautionary note and locates home confinement within its broader historical, social, and legal context. A balanced and informative introduction to the expanding practice of 'house arrest'."
—Gary T. Marx, *Massachusetts Institute of Technology*

Studies in Crime, Law, and Justice, Volume 3
1987 (Summer) / 236 pages (tent.) / $29.95 (c) / $14.95 (p)

SERIAL MURDER
by **RONALD M. HOLMES & JAMES DeBURGER**,
both at University of Louisville

This first systematic examination of serial murder takes a historical and theoretical approach to provide a thorough and practical guide to this phenomenon. The authors provide an overview of the cultural background, the social context, and the characteristics of serial murder and its perpetrators. Based on their own extensive research and on personal interviews with convicted serial murderers, this original work gives us new insights into this unexplored crime.

Studies in Crime, Law, and Justice, Volume 2
1987 (Summer) / 160 pages (tent.) / $26.00 (c) / $12.95 (p)

CRITICAL ISSUES IN CRIMINOLOGY AND CRIMINAL JUSTICE
by **JOSEPH E. SCOTT**, *Ohio State University*
& **TRAVIS HIRSCHI**, *University of Arizona*

Controversies in criminal justice are often framed in a "pro and con" format, emphasizing irreconcilable differences. The distinguished contributors to this original collection explore controversial issues in various parts of the criminal justice system, taking the approach that controversy should lead to recognition of common interests, and emphasizing a search for solutions. They examine various parts of the criminal justice system, beginning with types of crime, then focusing on the police, the courts, and finally imprisonment and its alternatives.

Studies in Crime, Law, and Justice, Volume 1
1987 (Summer) / 220 pages (tent.) / $29.95 (c) / $14.95 (p)

SAGE PUBLICATIONS, INC.
2111 West Hillcrest Drive,
Newbury Park, California 91320

SAGE PUBLICATIONS, INC.
275 South Beverly Drive,
Beverly Hills, California 90212

SAGE PUBLICATIONS LTD
28 Banner Street,
London EC1Y 8QE, England

SAGE PUBLICATIONS INDIA PVT LTD
M-32 Market, Greater Kailash I,
New Delhi 110 048 India

Greenwood Press and Praeger Publishers...
Where Social Theory Meets Practical Issues

CREATING THE WELFARE STATE
The Political Economy
of Twentieth-Century Reform
Second Edition—Revised and Expanded
by **Edward Berkowitz** and **Kim McQuaid**
Investigates how private business and public bureaucracy have worked together to create the structure of much of the modern American welfare state.
1988. ISBN 0-275-92747-4. $39.95

MANAGING PUBLIC LANDS IN THE PUBLIC INTEREST
Edited by **Benjamin C. Dysart III** and **Marion Clawson**
Discusses the controversies surrounding public land management including the impossibility of terminating public land use; the necessity of continuing private and multiple use; the need for sound policies to ensure the land's productivity; and the need for public involvement in land management.
1988. ISBN 0-275-92990-6. $39.95

SURVIVAL OF THE BLACK FAMILY
The Institutional Impact of American Social Policy
by **K. Sue Jewell**
Critically examines the social policies that arose from the civil rights movement and proposes new steps to economic independence for black families. October 1988. ISBN 0-275-92985-X. $36.00 tent.

HAS FREEDOM A FUTURE?
by **Adolph Lowe**. The Convergence Series. Founded, Planned, and Edited by **Ruth Nanda Anshen**.
"Adolph Lowe's book on freedom is a tract for our times, the wise testimony of one of the economists worthy of being called a worldly philosopher. It is a tough-minded analysis of the conditions needed for freedom in the modern age. I recommend it for everyone who is not afraid to wrestle with the future."—Robert Heilbroner, Professor of Economics, The New School for Social Research. 1988.
ISBN 0-275-92937-X. $35.95 (hardbound)
ISBN 0-275-92938-8. $12.95 (paperback)

POLICY STUDIES
Integration and Evaluation
by **Stuart S. Nagel**
Integrates the basic ideas that relate to policy studies, evaluates the methods of policy evaluation themselves, and assesses the field as a whole.
October 1988. ISBN 0-313-26256-X. $40.00 tent. (hardbound)/
ISBN 0-275-93007-6. $15.00 tent. (paperback)

AMERICAN CULTURAL PLURALISM AND LAW
by **Jill Norgren** and **Serena Nanda**
Examines the interaction of law with cultural pluralism in the United States, specifically the continual negotiation that has occurred between culturally different groups and the larger society. 1988.
ISBN 0-275-92695-8. $45.00 (hardbound)/
ISBN 0-275-92696-6. $16.95 (paperback)

SOCIAL GOALS AND EDUCATIONAL REFORM
American Schools in the Twentieth Century
edited by **Charles V. Willie** and **Inabeth Miller**
Provides a multidisciplinary perspective on the development of educational policy and practice in the United States during the past century.
1988. ISBN 0-313-24781-1. $37.95

Greenwood Press/Praeger Publishers

Divisions of Greenwood Press, Inc.
88 Post Road West, P.O. Box 5007, Westport, CT 06881, (203) 226-3571

Here's one reason why you need more life insurance...and three reasons why it should be our group insurance.

Family responsibilities increase and change—a new baby, a job change, a new home. Your family could have a lot to lose <u>unless</u> your insurance keeps pace with these changes.

Now here's why you need <u>our</u> group term life insurance.

First, it's low-cost. Unlike everything else, life rates have <u>gone down</u> over the past 20 years. And, because of our buying power, our group rates are low.

Second, you will continue to receive this protection even if you change jobs, as long as you remain a member and pay the premiums when due.

Third, our wide range of coverage allows you to choose the insurance that's right for you. And you can protect yourself and your entire family.

It's insurance as you need it. So check your current insurance portfolio. Then call or write the Administrator for the extra protection you need.

UP TO $240,000 IN TERM LIFE INSURANCE PROTECTION IS AVAILABLE TO AAPSS MEMBERS.

Plus these other group insurance plans:
Major Medical Expense Insurance
Excess Major Medical
In-Hospital Insurance
High-Limit Accident Insurance
Medicare Supplement

The AAPSS Life Plan is underwritten by New York Life Insurance Company, New York, New York 10010 on form number GMR.

**Contact Administrator,
AAPSS Group Insurance Program**
Smith-Sternau Organization, Inc
1255 23rd Street, N.W.
Washington, D.C. 20037

800 424-9883 Toll Free
in Washington, D.C. area, 202 296-8030